The Art and Science of Inventing

The Art and Science of Inventing

Second Edition

Gilbert Kivenson

VNR VAN NOSTRAND REINHOLD COMPANY
NEW YORK CINCINNATI TORONTO LONDON MELBOURNE

Manufactured in the United States of America

Published by Van Nostrand Reinhold Company Inc.
135 West 50th Street, New York, N.Y. 10020

Van Nostrand Reinhold Limited
1410 Birchmount Road
Scarborough, Ontario MIP 2E7, Canada

Van Nostrand Reinhold Australia Pty. Ltd.
17 Queen Street
Mitcham, Victoria 3132, Australia

Van Nostrand Reinhold Company Limited
Molly Millars Lane
Workingham, Berkshire, England

15 14 13 12 11 10 9 8 7 6 5 4 3 2 1

Library of Congress Cataloging in Publication Data

Kivenson, Gilbert.
    The art and science of inventing.

    Includes index.
    1. Inventions. I. Title.
T212.K58   1982   600   81-21799
ISBN 0-442-24583-1   AACR2

Preface to First Edition

Much of the material in this book is drawn from lecture notes used in a course on inventions and inventing. The course was presented at the Experimental College of the University of California at Los Angeles in the fall semester of 1974. The notes themselves were based on 20 years of industrial research experience and involvement in more than 50 developmental projects.

The backgrounds of students in the course were quite varied. Some were skilled industrial technicians. A few had relatively little knowledge of basic engineering principles, while others were doctoral candidates in the physical sciences. The reader will therefore find a somewhat wider than usual fluctuation in technical level to allow for as wide a range of previous study as possible.

The author would like to acknowledge the aid of Mr. R.F. Stengel, regional editor of Design News magazine, who read the manuscript and made helpful comments; Mr. Robert W. Kastenmeier of the U.S. House of Representatives, who supplied copies of pending patent legislation; and Assemblyman Terry Goggin of the California Legislature, who furnished a copy of his bill on inventor's rights. Reproductions of patents, indices, and other material were made from official documents of the United States Patent Office.

Canoga Park   GILBERT KIVENSON

Preface to Second Edition

In the four years since the publication of the first edition of *The Art and Science of Inventing*, the author has taught several courses in inventing technique at the adult-education level and one for an undergraduate engineering group. From questions and comments by the students, it appeared that the book could profitably be expanded, particularly with respect to practical examples and illustrations.

This edition contains two new chapters. The first is a study of current inventions in terms of their underlying principles. The second presents specific cases of profitable commercialization of inventions by independent inventors.

In four years there have also been significant advances in patent law and practice, e.g., the inventor's rights bills passed by Minnesota, California, Washington and North Carolina and the creation of the European Patent Organization. This edition has been updated to include some details of the important new legislation.

The author would like to acknowledge the aid of the following individuals and companies in providing material for this second edition:

David Benedict — Chatsworth, Ca.
Correlation Music Industries — La Canada, Ca.
Fusion Systems Corp. — Rockville, Md.
Gravity Dynamics Corp. — Los Angeles, Ca.
Joki Div., East Asiatic Co. — Greenboro, N.C.
Kobe, Inc. — Huntington Beach, Ca.
Polaroid Corp. — Cambridge, Mass.
Saray Engineering Co. — Wheaton, Ill.
Steven Titcomb — Stoneham, Mass.

G. K.

Contents

Preface for the First Edition   v
Preface for the Second Edition   vii

Introduction   ˙1

     The Independent Inventor   1
     Can Inventing Be Taught?   3

1. Basic Ideas About Inventing   4

     What Is Invention?   4
     What Is Not an Invention?   9
     The Areas of Invention   13
     Kinds of Invention   14

2. Some Fundamental Principles of Technology   21

     Energy and Its Forms   22
     The Transfer of Energy and Material   27
     The Storage of Energy   29
     Order and Chaos   31
     Some General Concepts   32
     Associated Art Effect   41

3. The Anatomy of a Patent   42

     The Patent   42
     Structure of the Patent   43
     Patents As Technical Literature   56
     How To Conduct a Patent Search   63

Libraries which Contain Complete Sets of Patents   67
Other Sources of Information   68

4. The Inventive Process   **70**

Preliminary Considerations   71
Problem Definition and Problem Assignment   73
The Enhancement of Creativity   75
Matrix Methods   78
Changing Viewpoints Method   80
Question Asking Methods   82
Updating and Adaptation   83
Biological Modeling   84
Analysis and Synthesis   87
Group Methods   93
Warm Up Methods   95
Pitfalls   97

5. Underlying Principles of Some Recent Developments   **99**

Motorized Drum — Combination of Two Functions   99
Plug-In Switch — Problem Analysis Followed by
   Specific Solution   100
One Piece Racquet — List of Shortcomings of Present Art   102
Ultrasonic Ranging — Biological Modeling   105
Roto-Jet® Pump — Function Reversal   107
Push Pedal Bicycle — Listing, Analysis, Association   111
Holography and Human Memory — Adaptation and
   Cross Fertilization   113

6. Planning the Experiment   **115**

General Principles of Experiment Planning   115
Steps in Experiment Layout   117
Simplification   124
Chemical Processes and Compositions of Matter   128
Overlooked Primary Factors   134
Summary   135

7. Apparatus Construction, Measurements and Data Handling   **136**

Introduction   136
Sketching and Drawing   137
Model Making   140
Measurement Taking   146

Transduction   154
Measurement System Design   157
Data Handling   162

8. The Psychology of Invention   167

Introduction   167
Motivation   167
Inner Needs   169
Maturity   172
Thought Processes   172

9. Obtaining a Patent   175

Introduction   175
Patent Services   177
Costs   179
Working with an Attorney or Agent   181
The Patenting Procedure   182

10. Making Your Invention Pay   186

Basic Considerations   186
Exploring the Possibilities   187
The Patented Idea   189
Locating Buyers   193
Patent Development Companies and Brokers   195
Manufacturing the Invention   198
Miscellaneous Considerations   201

11. Invention Into Enterprise — Some Case Histories   205

Fusion Systems Corporation   205
Correlation Music Industries   209
Dave Benedict Crossbows   213
Gravity Dynamics Corporation   217

12. Legislative Changes and the Inventor   222

Shortcomings of the System   222
Proposed Legislative Changes   225
European Patent Organization   231
Recent Court Decisions   232

Index   235

The Art
and Science
of
Inventing

Introduction

The Independent Inventor

Most persons have at some time in their lives invented something — perhaps a gadget for making some task easier or a method for increasing the economy of an everyday operation. Aside from a certain degree of personal satisfaction, the great majority of these inventions have never yielded their creators any amount of real return. In many instances, the fault has been with the inventor himself, who made little or no effort beyond writing down the bright idea. In a significant number of cases, however, the problem has arisen from a general lack of knowledge of what to do about a promising idea. The individual who works full time in a non-technical job usually has no guidance for proceeding in a logical, professional way towards effective development and utilization of his invention.

Several other factors are responsible for the considerable waste we see in the handling of inventions. Perhaps the most significant of these factors is an incomplete awareness of prior art. A stenographer dreams up a device to facilitate the distribution of incoming mail. She feels that the idea is so simple that others must have patented it long ago and so she goes no further with the concept. At the other extreme is the garage mechanic who invents a new type of wrench. He pours hard-earned funds into a patent application, only to discover that his ideas have indeed been anticipated by the patents of others. In both cases, the problems stemmed from lack of information and from not knowing how to obtain the necessary information.

A second factor responsible for much of the waste in the inventing process is the widespread belief that the era of the individual inventor is long past. The complexities of modern technology seem to indicate that inventing had best be left to professionals. Conclusions of this kind are incorrect and not based on fact. It is true that many developments are the result of close cooperation between experts of many disciplines, and costly apparatus is often employed to facilitate the search for a new device. Nevertheless, the basic concept behind each invention is still the product of an individual's imagination. Elaborate equipment is useful mainly for the accumulation of accurate experimental data. Very few, if any, true inventions are merely the result of the interpretation of data.

Although the equipment and physical plant of a large research organization are not needed for individual achievement in inventing, the methodology and scientific philosophy used by these organizations will be of great use to the casual inventor. If an inventor learns to use the professional approach, it will encourage him to continue with his invention, will suggest logical avenues along which to proceed, will prevent him from straying into fascinating but wasteful side efforts, and will help him to conclude each project in a satisfying and businesslike manner.

The proliferation of invention marketing "services" has not helped the cause of the independent inventor. Many of these organizations are interested primarily in the money to be made directly from the inventor's pocket — not from commissions to be earned by selling the invention to industry. One or two encounters with such organizations is enough to sour an inventor on the entire process and cause him to shelve all future ideas as fast as they are conceived.

A problem associated with inventing (as well as with many other endeavors) is that of communication. The independent inventor is often unable to describe his ideas and designs clearly to those in the best position to help him. His communications may not be technical enough to pass the hard scrutiny of a corporate research director and, at the same time, may be too technical for a small business loans administrator. The inventor needs guidance in modifying his presentation to fit either type of audience. He can also benefit from some auxiliary communications skills — e.g., a knowledge of perspective drawing or the ability to read and interpret the patent literature.

Can Inventing Be Taught?

Inventing, like art, can be taught. The mental mechanisms for increasing personal creativity can be stimulated. Undisciplined but highly innovative thought processes can be systematically channelled to produce streams of problem-solving concepts. Skills useful to the invention developing process can be learned. It is with these goals in mind that this book has been prepared.

1

Basic Ideas About Inventing

What Is Invention?

Before we begin any discussion about the technique of inventing, it would be profitable to think about what is and what is not an invention. A satisfactory set of criteria for use in defining true inventions is employed by the U.S. Patent Office. The term "satisfactory" is apt here, because the emphasis in this book will be on commercial practicality and industrial utilization of the invention. These are end results which distribute the work of the individual inventor for the greatest benefit of the general public.

The Patent Office rules state that the following types of innovation may be patented:

a. Processes
b. Machines
c. Methods of manufacture
d. Compositions of matter
e. Asexually-produced, living plants
f. Unique designs

Inventions not falling into one of these categories cannot ordinarily be patented. To better define the above classifications, let us illustrate each one with a specific example.

A process patent might be issued to an inventor who developed a technique for electroplating a plastic object to give it the appearance or utility of metal.   The process might consist of the following steps (see Fig. 1-1):

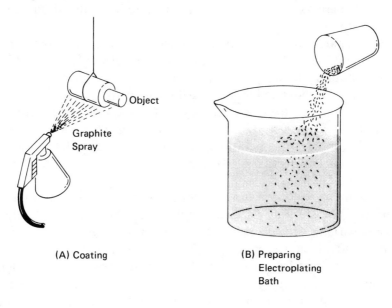

(A) Coating

(B) Preparing
Electroplating
Bath

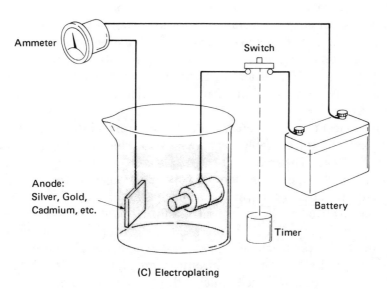

(C) Electroplating

Fig. 1-1. Process for coating plastic objects with various metals.

1. Coating the plastic object with a graphite paste to make it electrically conducting.
2. Preparation of a suitable bath containing dissolved metal salts.
3. Preparation of an anode, immersing it in the bath, and coupling it to one side of a direct current power supply.
4. Coupling the graphite-coated object to the other side of the power supply and immersing it in the bath.
5. Turning on the power supply and allowing a certain current to flow at a certain voltage for a certain period of time.
6. Finishing the operation. Removal of the plated object, rinsing, protection of the coating, etc.

There are several interesting properties associated with this kind of invention. In the first place, the individual steps can be operations well known and widely practiced; the invention may lie in the mere arrangement of the various steps. On the other hand, the newness of the discovery may be in the amount of current passed and the length of time over which it is applied. Another possibility of innovation is in the physical arrangement of the bath — e.g., there may be circulators in the solution so that no depleted regions are created, or the plastic object may be rotated to insure even distribution of the coating. These features may all be part of the process the inventor claims as something which is new and useful. (Even the recipe for the bath itself, if unique, is patentable, but it would properly be classified as a composition of matter, which we will be discussing shortly.)

The concept of a machine needs only a little elaboration. A machine is an assembly of operably connected parts which is used to produce a useful effect. The amplification of manual effort, the forming of raw material, and the measurement of some phenomenon are among the many useful effects for which machines are of value. The invention may be involved with the overall machine or with a small improvement in one part of it. The innovation might deal with an entirely new use for an old machine. A practical example of a highly effective machine is the bicycle. An invention might be concerned with a radically new kind of bicycle which is driven by a hand lever, or it could relate to improvements in speed shifting of conventional bicycles.

A method of manufacture, like a process, may be composed of a discrete number of steps taken in some definite order. Without this order or without all of the steps, it would be difficult, less-economical, or impossible to achieve the desired result. Methods of manufacture are usually defined in patent practice as acts used to physically prepare a raw material for use by a consumer. Processes, on the other hand, embrace a much wider interpretation and can apply to chemical, metallurgical, and nuclear treatments, among others. A hypothetical example of a method of manufacture which might be the subject of a patent is shown in Fig. 1-2. The item is a wall shelf sold in kit form to be assembled and painted by the purchaser. It is presently made by sawing out individual components several at a time (i.e., several sheets of wood are clamped together, the pattern is traced on the top sheet, and the sheets are band-sawed to shape). The cut surfaces are then sanded prior to packaging. The improvement consists of forming two thick blocks as shown in Figs. 1-2B and 1-2C. These blocks are shaped on an automatically controlled milling machine. Each of the thick blocks is repeatedly fed through a fine tooth circular saw which produces the individual shelf and end pieces. The surface finish produced by the milling machine and circular saw are sufficiently fine so that little or no sanding is required. As was the case with the process patent described above, conventional and well known steps (or equipment) may be used; the invention lies in the specific procedure taken to achieve a desirable result — in this case, higher production rates, a more uniform product, and lower labor costs.

A composition of matter invention deals with the forming of new and useful substances by the bringing together of several ingredients. When two or more materials are mixed, they may dissolve in one another, they may react chemically, or they may form various kinds of "alliances" — dispersions, colloids, alloys, suspensions, etc., or combinations of these. It is almost a certainty that the final substance will have different properties than any of the ingredients. In the simplest case, the product will have characteristics which are proportionally related to the quantities of each ingredient. In many instances, new and unexpected compounds, having drastically different characteristics than any of the original components, will be formed. The composition of matter patent may instruct the user as

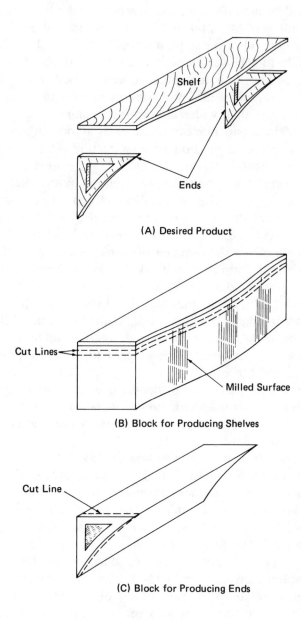

(A) Desired Product

(B) Block for Producing Shelves

(C) Block for Producing Ends

Fig. 1-2. Method of manufacture.

to the various ingredients he must employ, the quantities of each, the order of mixing, the temperature and pressure to be applied, the aging required, etc. An example of a simple composition of matter improvement would be the following recipe for an aftershave lotion:

	% by weight
Ethyl alcohol	70
Bay rum	15
Perfume	5
Coloring	0.5
Glycerine	3.5
Water	5
Alum	1
	100.0

In this recipe, the alcohol and bay rum are volatile cleansers to help wash away excess shaving cream; coloring contributes to the product's sales appeal; the glycerine makes the face feel smooth to the touch after the alcohol and bay rum have evaporated; while the perfume leaves a lingering fragrance. The improvement is in the addition of alum to the mixture; this imparts an astringent ("drawing together") property to the lotion so that small cuts can be made to stop bleeding without the additional use of a styptic pencil. Alum will not easily dissolve in the other ingredients without excessive stirring, however. The instructions for the preparation of the lotion, therefore, stipulate that the alum be dissolved in five times its own weight of water and the solution so obtained be then added in proper proportion to the rest of the lotion. This procedure (which may have been discovered by experiment) constitutes a part of the innovation.

The remaining two classes of invention, living plants and designs, are usually of minor importance to the independent inventor. We shall consequently dwell only briefly on these classes in later sections.

What Is Not an Invention?

The most generally known of the patent laws regarding the patentability of an improvement deals with its uniqueness. If the improvement conceived by inventor A was previously conceived by inventor B, who took steps to exploit his idea — patented it, manufactured it, sold it or published an account of its construction in some public journal — then the concept is *not* an invention as far as inventor A is concerned. Suppose, however, that inventor B has not acted on his advantage of prior discovery? There are basic rules which define the maximum periods of time over which inventor B is allowed to just "sit on" his idea before inventor A (who believes himself to be the original inventor) can obtain a valid patent. After maximum time has elapsed, the improvement now becomes a "non-invention" as far as inventor B is concerned. There are other rules which spell out procedure for cases in which two individuals apparently have the same idea at exactly the same time and both take steps to exploit the concept. Many hairline interpretations of the uniqueness rule have been necessary to treat fairly the special cases which have arisen when the same idea occurs within a short time to two or more inventors.

For an invention to be patentable, its originator must act in a diligent manner to put it in practical form, to apply for a patent as soon as possible, to keep its details secret from the public, and to act, in general, as though he means business. The lackadaisical inventor who allows a brilliant idea to languish on some note paper tucked away in a drawer, stands to lose his legal right to patent if another, more aggressive individual comes along with the same idea.

Any process, machine, method of manufacture, or composition of matter, regardless of its uniqueness, is not an invention if, in its intended use, it will damage public well-being or morality. Thus, a gambling machine, a counterfeiting process, or a method for growing marijuana undetectably cannot be patented. This rule does not apply to weapons which are useful in war even though tremendous harm can be done to large numbers of individuals with these inventions. If the invention is judged by the Patent Office to have special strategic value (e.g., an atomic weapon) it will be first submitted to a government agency. If the latter desires the matter to be kept secret, the normal patenting process is held in abeyance and the invention essentially becomes the property of the United States government. The inventor will be directly compensated at some later date by the agency which utilizes his idea.

Several other kinds of "non-inventions" are of interest here because they are frequently created by independent inventors. The Patent

Office does not consider a new method of doing business as an invention even though it meets the requirements of being novel, unique, useful, never before revealed, etc. If, for example, the innovation concerns a new system for inventory control using a special and simplified data sheet, no patent protection can be obtained on the system itself. It is possible, however, to patent any special apparatus used with the data sheet (for instance, an electronic pencil which both reports through a computer that a certain item is getting low, and is also used to mark up the sheet). The data sheet itself can possibly be copyrighted to protect its form or literary content.

A pure scientific principle cannot be patented. If the inventor discovers a significant relationship between the color of a soil and the maximum number of potatoes which will flourish in it, he cannot protect this principle as such. If he devises a meter for measuring the color of the soil and thereby predict its potential productivity, the meter falls within the class of patentable matter.

The Patent Office will not consider as invention a naturally occurring item even though it has never been found before by anyone else. The invention must consist of some substantial alteration (in this case) of the item to make it more useful. If the living plant type of patent is to be obtained, for example, the inventor must show that he has asexually produced a new species of flower, bush, tree, or vegetable from an existing species by grafting, layering, rooting of cuttings, or some other method.

A beginning inventor will often not think of working out the details of his concept. Patent attorneys, as well as the new product departments of corporations, will often receive letters outlining a scheme for some improvement but not specifying exactly how the idea is to be implemented. As an example, an inventor may propose harnessing the coastal tides to produce electricity. He may have partly thought out the problem — to the extent of suggesting stand-by fossil fuel generators to supply power during periods when tidal movement is at a minimum, so as to assure a constant supply. What is missing, however, is a practical embodiment for converting the motion of many tons of widely distributed sea water into a concentrated force capable of turning a small number of generators. The individual who submits this type of proposal will, of course, be turned down; he often feels cheated at some later date when he reads about "his" scheme being put into successful operation. What has happened is that the missing but vital portion of the concept was provided by an arduous research effort on the part of some corporation. The research was not necessarily inspired by the inventor's original suggestion; the problem may have been encountered in the course of everyday business.

The final class of "non-invention" to be discussed here is the device or concept which is "apparent to those skilled in the art." An example will make this clear. A beginning inventor will often re-invent the toothpaste tube squeezer. The concept is usually reduced to practice in the form of a holder (enclosed or open) which attaches to a wall. The toothpaste is clamped into the holder where it is subject to the uniform pressure of a metal bar or roller; the paste is collected at the bottom directly on the brush (Fig. 1-3). Aside from the fact that this improvement is periodically re-invented (which would be a bar

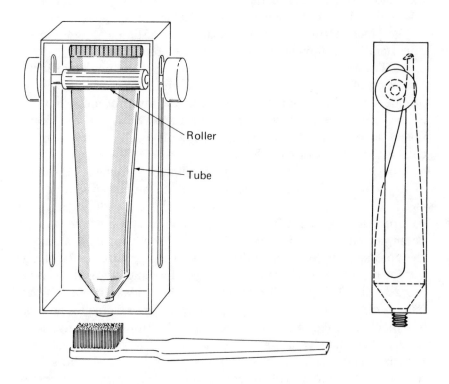

Fig. 1-3. Toothpaste dispenser.

to patentability on the grounds of its not being an original idea), the concept would still not be a true invention. Any person skilled in the design of household appliances could readily concoct a similar or equivalent device if he were to apply himself to the problem of getting the entire contents out of a squeeze tube. It should be mentioned here that some qualitative judgment is involved in deciding what is apparent to a skilled practitioner and what is not. This judgment has been the subject of much patent litigation.

The Areas of Invention

There are 16 general areas which are served by most patents. The classifications and subdivisions used below cover a vast majority of inventive effort. The areas are:

1. Raw material treatment: mining, enriching, refining, recovery.
2. Manufacturing: components, assemblies, consumer products, industrial products.
3. Construction: large structures, dwellings, urban layouts, roads.
4. Transportation: vehicles, navigation, traffic control.
5. Communications: transmitting, relaying, receiving, distribution.
6. Power generation: production, distribution.
7. Farming: soil preparation, growing, harvesting, maintenance.
8. Medical: pharmaceuticals, apparatus, systems.
9. Commercial fishing: equipment, treatments, bait.
10. Food preparation: preserving, cooking.
11. Military: logistics, weaponry, systems.
12. Home: appliances, fixtures, furnishings, maintenance, comfort.
13. Office: appliances, supplies, maintenance.
14. Toys: games, sports, equipment, systems.
15. Personal: clothing, cosmetics, maintenance.
16. Entertainment: public, home.

It is often surprising to the seasoned inventor, when he looks back over a considerable number of his developments, to find that they have been confined to a relatively small area. Although narrow and intense specialization does improve the quality of innovation, other areas which could benefit from an identical invention are often overlooked.

Kinds of Invention

If we were to study a large number of inventions, speculate on how they had been conceived, and then tabulate the results, we would find that a relatively small number of "mental mechanisms" had been used. This does not detract in any way from the ingenuity or quality of the thinking brought to bear by the individual inventors. It merely tells us that a methodology can be derived which may be useful in obtaining ideas more quickly and in evaluating them afterwards. There are, of course, some highly creative people, the inner workings of whose minds are mysteries even to their possessors. Evaluating the inventions of these individuals as to the probable system used is a rather speculative process. In most cases, though, inventions are found to be classifiable under one of the following types:

1. *The single or multiple combination.* The most elementary form of invention is a simple combination of two already existing devices in order to achieve an improved result. A pen and pencil on opposite sides of the same holder create certain properties not possessed by either item alone. The user can conveniently change from pencil to ink without having to carry an extra writing utensil in his pocket. The pull-along golf cart which embodies a small writing table, means for holding the score card, and a pull-down seat for resting, is an example of a combination in which utility is endowed by the act of combining four separate devices. There are, however, some pitfalls in the use of simple combination as a mode of inventing. The patentability of some combinations is often questionable. There is first the previously discussed rule which eliminates innovations which are apparent to those skilled in the art. The Patent Office also disapproves of concepts representing "mere aggregations" of individual items which are not operably related. A dial telephone combined with a washing machine, for example, would be such an aggregation. On the other hand, a drill press, lathe, and circular saw combination would be a true invention because the apparatus permits

the use of a common motor-pulley system, and parts of the drill press are readily and conveniently transferred to become the lathe or circular saw. The latter combination possesses several valuable and unique properties — it is less expensive than three separate machines would be; it requires less room to operate, since only one machine is assembled in any given operation; and it can be put into a "standby" stage and compactly stored. In this case, the design of a combination which permits easy assembly and conversion from one function to another represents inventive ability greater than that possessed by someone with ordinary skill in the art.

An interesting exercise in creativity which can lead to many inventions is to choose two pages at random from a large mail order catalogue. Select one item or object from each page and try to combine the chosen items into an innovation which is useful and unique. In one of the author's classes, the items chosen from a department store catalogue were a camping tent and a roller type window shade. In a cooperative effort, the class blended the functions of both items. The result consisted of a skeleton frame, two members of which contained rolled up tent canvas. The tent skeleton frame could be readily bolted together as shown in Fig. 1-4. The covering operation would then only require the canvas to be fed out from the two rollers and secured to the frame by means of sewed-on ties.

2. *Labor saving concepts.* A somewhat higher degree of sophistication in inventing is displayed when we modify some existing assembly technique or process in order to save effort, produce more with the same effort, or achieve unattended operation. Whenever a new source of power is introduced, there is always a rash of inventions which couple the new source with existing devices. When the steam engine was perfected, it was applied in rapid succession to pumping water, sawing wood, and driving vehicles. The electric motor was no sooner developed than many inventions appeared which utilized it to replace the bulkier and less efficient steam engine. The properties of the electric motor are still being exploited today because we have learned to miniaturize it without losing high torque and efficiency. Using tiny but powerful electric motors, we have not only electrified the sewing machine and the broom but have added power to the razor, typewriter, and even the carving knife! In a similar manner,

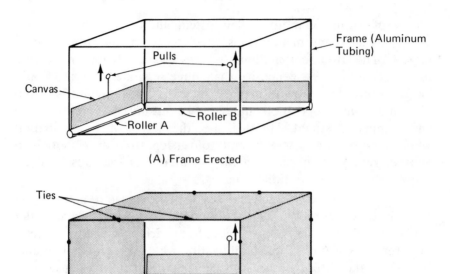

(A) Frame Erected

(B) Tent Partially Covered

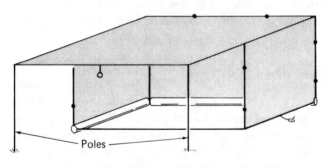

(C) Tent Completely Erected

Fig. 1-4. Invention by combination.

the miniaturized internal combustion engine has been applied to lawn mowers, chain saws, air compressors, and small boats. Again, however, the question of patentability arises. The addition of a motor to a machine which was formerly driven by hand does not constitute true invention unless some new advantage is gained thereby. Coupling an electric motor to a kitchen mixer and driving it at high speed permits the uniform addition of air to the food; this extends the use of the mixer and imparts a pleasant modification of taste not possible when the mixer is operated by hand. This machine, now called a blender, represented a patentable improvement over prior art. The coupling of an electric motor to a pencil sharpener, on the other hand, does not increase the speed or efficiency of the operation. Although some argument might be made here for the convenience of one handed usage, this innovation could not as readily be defined as true invention.

3. *Direct solution to a problem.* A type of invention which requires somewhat more analytical reasoning than the combination or labor-saving kind is the device which is "custom" conceived to solve a particular problem. Here the inventor does not perceive some new arrangement of parts and then attempt to find uses for the concept. He is instead confronted with a need and sets out purposefully to design a system which will fulfill that need. A hypothetical case will illustrate the point. A woman is preparing a soup which requires frequent stirring. When she puts the spoon down on the stove surface, a stain results. She attaches a hook to the spoon handle which permits the spoon to be hung on the inside of the pan when not in use. This is not simple aggregation because the two items, spoon and hook, are operably connected. The spoon-hook device proves impractical because the handle becomes too hot when the spoon remains in contact with the soup. She next adds an insulated handle but this also has shortcomings because of heat received directly from the flame. Eventually she devises a shield, fastened between the spoon and insulated handle, which permits the spoon to remain hooked to the pan indefinitely. The final invention, in this case, consists of three separate problem-solving stages applied to a definite problem.

4. *Adaptation of an old principle to an old problem to achieve a new result.* This is a variation of the problem-solving mode described

in the previous section. Additional sophistication, however, may be involved in the methods used to solve the problem. In this particular mode, the problem has been in existence for some time; the principle to be used in its solution is known. The uniqueness and creative innovation comes about in bringing this particular principle to bear on this particular problem so as to achieve a useful result. Two recent examples will clarify this method. Pencil leads have been made for many years from powdered graphite fired with clay; the clay serves to bond the graphite and to control the "softness" of the finished product. It was not possible to produce these leads in diameters much under one millimeter because of the brittleness of the fired clay. The substitution of plastic (a well known material) for clay permitted sturdy pencil leads to be made in 0.5 and 0.3 millimeter sizes. These leads have been widely sold to draftsmen and artists because they permit fine lines to be drawn without sharpening or sanding.

Another example is the ultrasonic burglar alarm. Ultrasonic vibration has been used for years in such diverse applications as homogenizing milk and cleaning jewelry. The application of an ultrasonic field to an enclosed space permits the detection of any slight movement in that space because the wave pattern is readily disturbed and the disturbance can be picked up with a special microphone.

5. *Application of a new principle to an old problem.* Sometimes an existing problem has been only partially solved by existing techniques. The inventor who takes advantage of some new technology can often achieve a resounding success. An example was the application of transistors to hearing aid technology. The vacuum tubes previously used were bulky, fragile, and required too much power. First the transistor and then the integrated circuit permitted the hearing aid to be greatly miniaturized and to operate for long periods from a tiny battery.

Another example of a new principle applied to an old problem is the recovery of petroleum by the use of underground nuclear explosions. It is possible, with this technique, to create, at great depths, huge spherical enlargements into which oil from surrounding sands will seep and accumulate.

6. *Application of a new principle to a new use.* New problems are continuously being recognized by inventors. Those who have

knowledge of the latest techniques and technologies are in a position to apply one or more of these to satisfy a newly-created need. A recent example has been the deployment of communications satellites in stationary positions above the earth and the use of these line-of-sight stations to relay programming and communications around the world. The new problem was the increasingly heavy traffic and load on existing communications channels and the need for extra overland and submarine cables. The new principle used to satisfy this need was satellite-relaying of signals.

7. *Serendipity or the principle of exploiting lucky breaks.* Everyone is familiar with stories about accidental discoveries which led to great inventions. There seem to be two types of these strokes of good fortune. In the first kind, the inventor is actively engaged in problem-solving but is unable to go past a certain point in his progress. A freak occurrence or chance observation then provides the answer. In the second type, the inventor suddenly gains a valuable insight or discovers a new principle not related to the work in which he is engaged but to some other area. He then applies the discovery to the new area and, as a result, is highly successful. An example of the first type is the famous case of rubber vulcanization. Charles Goodyear had been experimenting with natural latex solutions but had been unable to stabilize any of them into a non-sticking, flexible mass which would completely recover after being compressed. He had tried sulfur as an additive, but only after the accidental spilling of this mixture onto a hot stove did he discover that heat treatment would produce the desired result. An example of the second type of good fortune would be that of the unknown individual who first discovered the explosive properties of a potassium nitrate, charcoal, and sulfur mixture. It does not seem likely that the experimenter was seeking a method for driving bullets at high speeds through gun barrels or for blasting rocks for road construction.

It should be noted here that in either type of accidental discovery, the individual concerned does not benefit unless he is receptive, is capable of deep analytical reasoning, and possesses sufficient flexibility to alter his approach in accordance with the direction shown by his findings. John Ericsson, the inventor of the marine propeller, had equipped a test vessel with an engine-driven appendage which resembled a large corkscrew. In his first large scale test, the vessel began to move under the thrust provided by this device, but hydraulic forces were too great for the metal which the screw was made of. The metal fractured and the screw broke off, leaving only a single convolution on the drive shaft, and the speed of the vessel increased sharply. The inventor had theorized that several screw threads were necessary to prevent slippage of the water away from the driving element; a fortunate accident (and its proper interpretation) taught him that a single thread, if rotated rapidly enough, gave both minimum slippage and efficient utilization of the engine power.

2

Some Fundamental Principles of Technology

One often-noted difference between part-time inventors and industrial research workers is the scientific training of the latter. The corporate scientist is well versed in technical principles so he can often eliminate certain approaches in attempting to solve problems. Some experiments which suggest themselves to an untrained person would require the "repeal" of well known laws of physics, chemistry, or engineering. The expert does, however, pay a price for his knowledge. The inventor without formal training will sometimes discover an unsuspected exception to a well-established rule and will thereby create an extremely valuable invention. An example of this occurred in the middle forties when a radically different refrigerating device, called the Hilsch tube, was revealed. The invention actually separated hot and cold molecules of air pumped through it. The cold molecules, when collected, were at a temperature substantially below room conditions and could be used for refrigerating purposes. Molecular separation of this kind had been believed to be in complete contradiction to a classical principle of thermodynamics.

Despite these exceptions, however, it is more generally the case that the untrained individual will follow many "pre-doomed" paths and, thereby, lose much valuable time. In order to minimize this tendency, we will use this chapter to discuss, in highly condensed form, the basic principles, from an inventor's standpoint, on which present day technology is based.

Although the description to follow can only be considered a bare outline of a vast area of knowledge, it will serve to familiarize the untrained inventor with general terms and fundamental laws, and it will permit him to undertake more detailed texts at a later time. For those who have studied the basic sciences, this description may serve as a brief review.

Energy and Its Forms

Whenever a force is applied to move a load, energy is expended. The numerical value of the force multiplied by the distance through which the load is moved is a measure of the energy used. It is always found that the total energy put into an operation is greater than that recovered as useful work. We speak in practical cases of a conversion efficiency; the latter is the useful work obtained divided by the input energy times 100 (conversion efficiency = $\frac{\text{energy out}}{\text{energy in}} \times 100$). In some types of machinery, the conversion efficiency is very high: e.g., 90%. In many others, it is much lower because of inherent characteristics of the process or because of lags in the development of certain arts.

There exist seven known forms of energy: mechanical, electrical, optical, chemical, acoustic, nuclear, and heat. Some overlap exists in these classifications; radiative heat transfer, for example, consists of optical transmission of heat energy. Acoustic energy transfer makes use of mechanical vibrations of a column of air.

It is possible to convert one form of energy into another with varying degrees of efficiency. Mechanical energy can be converted to electrical, electrical to optical, optical to heat, etc. A basic law of thermodynamics states that energy can neither be created nor destroyed; it can only be converted from one form to another. Figure 2-1 is a chart of the various commonly-made conversions. Mechanical work can be converted into electrical energy by means of an alternator, chemical energy into electrical by means of a battery, and optical energy into electrical by use of a photocell. It will be noted that some of the blocks in Fig. 2-1 are connected by opposite-going arrows, indicating that the conversion process is easily reversed. An arrow in one direction only indicates the "easy" path; reversal is difficult or may require more than one process. It is difficult, for

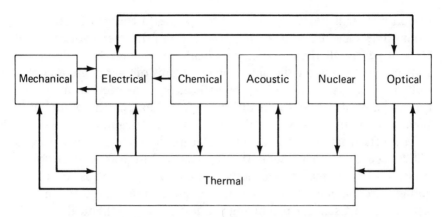

Fig. 2-1. The known forms of energy and their conversion.

example, to produce chemical energy directly from heat. Thermal means can, however, be used to refine chemicals from which batteries can be constructed. It is also noteworthy that six of the energy types can be converted easily to heat. In a sense, heat is the "lowest" form of energy and is the "rest state" to which other forms will revert. To obtain other types of energy from heat, it is necessary to pass the thermal energy through certain types of converters. Heat energy can only be transferred by the use of temperature differences. The converter extracts a certain percentage of the heat passing through it, changes this into the desired form of energy, and then rejects the unused heat at some lower temperature. No matter what temperature extremes are used to drive the heat into and out of the converter, some heat will always leave unconverted. Figure 2-2 is a block diagram indicating this characteristic of thermal converters.

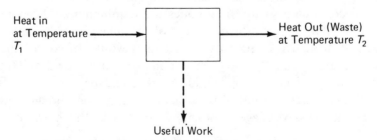

Fig. 2-2 The heat engine.

In contrast, it is perfectly feasible to transfer a charge of electricity from some elevated voltage to ground potential (zero voltage) and convert nearly all the electricity to heat. In general, high efficiency conversions among all the six "nobler" forms of energy are theoretically possible, and all six can be efficiently converted to heat. To produce any of the six forms of energy from heat, on the other hand, is an "uphill" process and is inherently of low efficiency.

The efficiency of any conversion can, of course, never exceed 100% because of unavoidable transfer losses. In the above-mentioned transfer of a charge of electricity, for example, some loss occurs in voltage drop along the wires which conduct the charge to any resistance load used for producing the heat.* In mechanical arrangements, there is always friction; in optical devices, there are always scattering and absorption; in chemical systems, there are always side reactions.

Patent applications in the past often dealt with perpetual motion. The concept usually proposed is shown in Fig. 2-3A. The scheme calls for an initial investment of energy E_1 to get the machine started. A converter provides a certain amount of useful work W_1 and rejects the rest of the energy E_2 at some lower potential. W_1 is used to operate a pump which restores E_2 to its initial high level (E_1) and recycles it back to the converter. The valve is closed once the machine has started so that no additional energy is added. The net useful work W_2 (the total work W_1 minus the small amount of work required to operate the pump) is then available for external purposes. A "practical" configuration of this principle is shown in Fig. 2-3B. A motor is directly coupled to an electrical generator and a starting current is applied to the motor through a switch. The motor and generator are brought up to speed and the output of the generator is then connected so that it supplies the requirements of the motor. Work is obtained from the other end of the generator shaft as shown. Practical difficulties with this arrangement would, of course, include the fact that even a "perfect" motor and generator could only match each other and no useful outside work could be obtained. In an actual case, there would also be heat produced in the windings of both machines, windage losses, and friction in the bearings. The

---

* At very low temperatures it is possible to reduce resistance to zero, but additional energy is required for refrigeration.

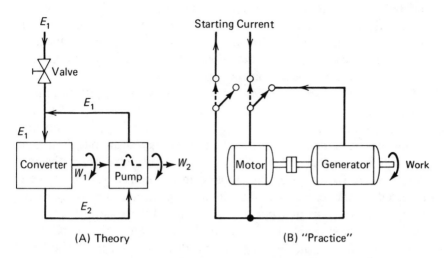

Fig. 2-3. The perpetual motion machine.

Patent Office does not generally require (as it once did) working models as a part of each application, but perpetual motion schemes are a significant exception. The inventor must demonstrate by a functioning device that he has indeed overcome the basic problems preventing the practical realization of this kind of machine. It must be pointed out here that machines which depend on waterfalls, atmospheric pressure changes, diurnal temperature variations, seasonal changes, etc. are not perpetual motion machines, even though they may operate for long periods of time without additional power.

Sometimes a power scheme is conceived in which the perpetual motion aspects are difficult to detect. A system that is periodically reinvented decomposes water into hydrogen and oxygen, two gases that can easily be recombined to produce intense heat, electricity (as in a fuel cell), or pressure increase (as in an internal combustion engine). If we propose to perform the decomposition by electrolysis (i.e., apply two electrodes to a quantity of water made conducting with acid or base, and collect the gases at each electrode), a relatively simple calculation will show that the input energy required will be equal to or greater than that which can be recovered. If, however, a chemical substance is employed to generate hydrogen and oxygen from water, an evaluation is often not as easy to make. It may, in fact, be quite difficult to assess the energy input required to prepare the chemical substance or to refine the ingredients so that they will be active when in contact with water. Despite any vagueness in these estimations, however, it can safely be assumed that the maximum output will always be less than the input.

An apparent exception to this relation between output and input occurs whenever a life process is involved. If, for example, we use bacteria to produce hydrogen, methane, or alcohol from organic substances, it appears that the output energy available from the combustion of these products exceeds that available from the raw materials. A factor often overlooked in the consideration of these processes is the capability that living organisms have of drawing energy from the sun or from media warmed by the sun. Plants and certain bacteria are able to use warm atmospheric gases in adding to their own bulk and in creating side products by cell multiplication, nitrogen fixation, etc. Therefore, a power source utilizing living

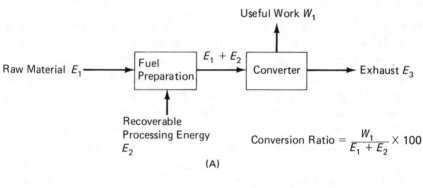

(A)

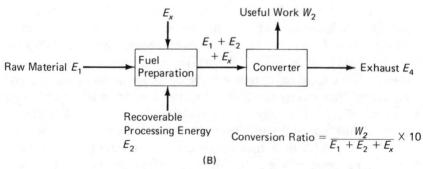

(B)

Fig. 2-4. Energy conversion processes.

processes can be considered in the same general category as water falls and windmills. Figure 2-4 illustrates schematically some of the principles discussed above. In the upper illustration is shown a process in which raw material having an initial energy E_1 is prepared into fuel by the addition of processing energy E_2. The finished material, now having a theoretical energy content of $E_1 + E_2$, is fed into an energy converter where W_1 units of useful work are obtained. The exhaust carries with it lost energy E_3. The conversion efficiency, as defined before, is the useful work divided by the total input energy times 100. Omitting either term in the denominator, E_1 or E_2, gives an erroneously high value of the conversion efficiency. Figure 2-4B shows a process in which additional external energy E_x is added in preparing the fuel. E_x could represent solar energy added by a biological process. Again, the omission of any term in the denominator when we calculate the conversion efficiency will lead to an erroneously high value, and possibly the impression that perpetual motion has been achieved.

The Transfer of Energy and Material

We have stated above that a driving force in the form of temperature difference is required to move heat energy through a converter in order to obtain useful work. It is found that all flow, whether it be of energy or of material, requires some form of driving force. The latter is referred to in technical literature as a "differential" — i.e., the difference between the upstream and downstream forces needed to move an object or quantity of energy in a given direction. The greater the differential, the faster will be the movement or flow. The size of the channel or passageway will also be important. The smaller the passageway, the lower will be the flow for a given differential. The general relationship is a simple one:

$$q = \frac{(F)^n}{r} \tag{1}$$

where q = the flow rate of energy or matter
F = the differential or driving force
r = the "resistance" of the passageway to flow
n = a constant

This fundamental law has a number of forms, depending on the field to which it is applied. Some of its well known forms and the names of the individuals given credit for them are listed below:

Electricity $\qquad\qquad I = \dfrac{E}{R}$ $\qquad$ Ohm's Law ($n = 1$) $\qquad$ (2)

> where I = rate of flow of electricity between two points in a wire
> E = voltage differential between the two points
> R = resistance of the wire to electricity between the two points

Heat $\qquad\qquad Q = \dfrac{T_h - T_c}{r_t}$ $\qquad$ Fourier's Law ($n = 1$) (3)

> where Q = rate of flow of heat
> $T_h - T_c$ = the temperature differential between hot and cold points in a solid through which heat is flowing
> r_t = "thermal" resistance of the solid between the two points

Fluid Flow $\qquad U = \dfrac{\sqrt{P}}{r_f} = \dfrac{(P)^{1/2}}{r_f}$ $\qquad$ Bernoulli's Law ($n = \frac{1}{2}$) (4)

> where U = rate of fluid flow between two points in a pipe or conduit
> P = the pressure differential between the two points
> r_f = the resistance to flow offered by the pipe or conduit

Equations (2), (3), and (4) apply to flows when rates are low to moderate. If we wish to increase the flows to higher values, other factors require consideration. At high fluid flows, for example, friction becomes more and more important and the equation describing the relationship between flow and pressure becomes more complex. At high heat transfer rates, conduction of thermal energy through solids is not the only transport mechanism; convection and radiation losses become increasingly significant. In the transmission of high amperages through conductors, minute impurities in the material contribute enough resistance (negligible at lower currents) to cause heating of the wire and consequent thermal losses. Many successful inventions have come about by the utilization of the fact that limiting laws change drastically if flows are greatly increased (or decreased). A number of years ago a steam boiler was invented which made use of extremely high water flow rates through its tubes. The water which was not evaporated on the first pass was rapidly separated from the steam and recycled through the heating chamber. At the very high rates used in this design, the resistance to heat transfer (r_t in Eq. (3)) decreases greatly. The boiler could come up to full output within five minutes of start-up and was very much smaller for a given capacity than any boiler previously designed.

The Storage of Energy

We have previously listed the seven forms of energy: mechanical, electrical, chemical, acoustic, nuclear, optical and thermal. Each of these exists in one of two basic conditions: potential or kinetic. A one pound weight resting on a shelf six feet above the floor represents a potential energy of six foot pounds. If the weight is moved over to the edge and allowed to fall to the floor, the six foot pounds is converted to mechanical work. The water impounded behind a dam is a large scale example of mechanical energy stored in potential form. If the water is conducted through pipes to turbines below, the energy becomes kinetic and is converted to electricity by generators coupled to the turbines. Placing energy in potential form represents one form of storage. It is, however, not the only method. A spinning flywheel also represents stored energy. An automobile can, in principle, be run by drawing from the energy stored in a spinning flywheel. The potential energy for the vehicle would be "recharged" by periodically connecting the slowed flywheel to a small gasoline motor in the car and again building up a high speed.

Energy storage is a continuing technical challenge because of the variable nature of most energy sources. The sun does not always shine, rivers do not always deliver the same amount of water, gas wells decrease in pressure as the supply diminishes. Energy must be stored and then metered out in uniform amounts or in amounts

TABLE 2-1. METHODS FOR STORING ENERGY.

Type	Methods
Mechanical	Water reservoirs, flywheels, springs
Electrical	Condensers, rechargeable batteries
Chemical	Fuels, reagents
Acoustic	Resonating chambers
Nuclear	Radioactive materials
Optical	Phosphors, glowing substances
Thermal	Heated objects, melted solids, condensed gases

proportional to fluctuating requirements. Some commonly used methods for the storage of each form of energy are listed in Table 2-1.

The sun is the primary source of all terrestrial energy. Solar evaporation and the subsequent condensation of water provide the energy found in stored bodies of water — rivers originating at higher elevations or from large masses of melting snow. Heating of air masses produces density changes, large scale movements, and the production of wind energy. Even the energy in atoms originated in the original planetary formation process which took place within the body of the sun.

A storage process still difficult to understand and not yet duplicated in the laboratory is that of photosynthesis. Atmospheric carbon dioxide is converted to carbohydrates by growing plants which use sunlight as the energy source and earth minerals and chlorophyll as catalysts; the carbohydrates can then be burned to release thermal energy. Many generations of vegetation have been buried by landshifts and earthquakes. Extreme pressures exerted at great depths have formed the oil and coal deposits which represent our most important source of stored energy. Present schemes for the utilization of solar energy on an industrial scale often do not allow for the storage time required if effective amounts of energy are to be obtained. The intended drain must be compared to the maximum obtainable input rate. If the drain is greater at any time during the day than the input, then some back-up energy source is required.

Order and Chaos

Much of the scientific thinking of the early part of the 19th century was concerned with the philosophical aspects of energy transfer. The concepts developed then are still of great importance, and a brief consideration of them will constitute good background material for the inventor. The 19th century theorists proposed that the ultimate "rest state" of all nature is complete chaos and disorder. It was known that atomic particles move randomly; non-elastic collisions between them will eventually cause all energy to become heat (the "lowest form" of energy). If it is further postulated that the universe is expanding, this heat will distribute itself over a greater and greater amount of space until the overall temperature reaches $-459.9°$ F or "absolute zero." At this point, all atomic motion will be stopped and energy will no longer exist. It was theorized that any processing of a material which halts or reverses its movement towards chaos requires an energy input. The natural tendency of any material so processed is to return to its original condition at an accelerated rate (and then resume its much slower deterioration towards complete randomness). From this standpoint, any manufacturing operation can be considered as an "arranging process." The randomness of the raw material is decreased by forming, machining, welding, etc. The finished product has a certain amount of stored energy represented by a molecular structure which has been displaced from its rest state. Under the stress of usage, the manufactured object will tend to liberate some of this stored energy and return to a more disordered state. Wear, corrosion, breakage, etc., are some of the mechanisms by which this return occurs. An inventor seeking to prolong the life of a manufactured object can often find a method for the safe dissipation of some of the stored energy or design a by-pass so that the liberation of stored energy does not interfere with the function of the object. An example of a stored energy source which acts destructively is the welding stress locked in when a weld cools. Heat treatment will safely reduce this stress. The weld, however, is more active chemically than the materials which have been joined and will tend to rust more rapidly. A spring-loaded bearing holder is an example of a design which tolerates deterioration with minimum interference with function. Under the stress of use, the bearing will be subject to erosion — a grinding process which tends to return the bearing to the raw material from which it was fabricated. A spring-loaded bearing holder, by constantly realigning the bearing pads, compensates for wear and prevents malfunctioning during the useful life of the pads. Many inventions are concerned with techniques for overcoming the natural tendency of all fabricated materials and assemblies to deteriorate and break down.

Some General Concepts

If a study is made of modern technology, it is found that a number of concepts are used over and over in originating, developing, and applying many devices and systems. Each of these concepts is not usually confined to a single scientific rule but incorporates several laws of physics, chemistry, or engineering. Consideration of these concepts will be valuable to the inventor because of the generalized and concise survey of present day science they afford. The concepts are:

1. *The area principle.* Penetration of one material by another depends not on the force applied but on the force per unit area. It is obvious that a large force spread over a large area will produce only minimal penetration. In the construction of a building, it is desirable to distribute the load over as wide an area as possible. In the design of a cutting edge, on the other hand, the area of contact is made as small as is consistent with the retention of cutter strength. An available force of 5 lb., when applied over an area of 1 in.^{2}, produces a pressure of 5 lb/in.^{2} The same force applied to a needle point having an area of .001 in. increases the pressure 1,000 times to 5,000 lb/in.^{2} All mechanical cutting processes depend on the use of this principle. Although the concept seems straightforward enough, its application is not without some problems. The high pressure exerted on a material to be penetrated is also exerted on the cutting tool. It is easy to exceed the compressive strength of the tool and cause collapse. The contact area is thus increased and the effectiveness of the tool is lost. It is then necessary to reform the contact area by resharpening. Other difficulties associated with this use of the area principle involve removal of material which is displaced by the penetration and the dissipation of heat produced.

2. *The feedback principle.* This concept involves the monitoring of an operation to check for proper output and the continuous adjustment of operating variables to maintain the process at some preset condition. The feedback principle is modelled directly after inherent human behavior and attempts to duplicate the complex responses of the body when carrying out a task. The driving of a car, for example, incorporates feedback from visual, auditory, and position signals. The muscles of the arms and legs are triggered by these signals and act to maintain the vehicle in the proper lane and at constant speed despite a non-planar, wandering road, paved with varying types of surfacing. Industrial control systems, vehicle and ship navigation devices, environment regulators (e.g., heating and air conditioning), and many other systems make use of the feedback principle.

3. *The large and small principle.* The relative strengths of large and small creatures have intrigued men for many centuries. An ant can easily lift an object weighing six times its own weight and survive a fall from any height. A toy ship or car can withstand collision forces which, if brought up to scale, would completely destroy a full-sized version. It is found that strength, structural resistance, flexibility, and other properties of materials vary in a non-linear manner with dimensions. The building of a model of a proposed larger structure and the obtaining of results which can be scaled up requires careful mathematical analysis. Proper dimensioning of the model involves the arrangement of the numbers representing size, velocity, friction, and other applicable variables into certain mathematical groups which can be correlated with overall performance.

Small objects have greater relative strength than large ones for a number of other reasons. In the cooling of a melt of steel, for example, there is uneven shrinkage, formation of gas pockets, and aggregation of impurities at certain points — all of which tend to weaken the resulting casting. A fine wire drawn from this same steel will, on the other hand, be much more uniform. As the original material is heated and formed into rods which can then be fed to a wire drawing machine, many of the gas pockets and impurity concentrations can be eliminated. The wire consequently has greater tensile strength than the original steel casting. In recent years, experiments have been made with steel "whiskers" drawn from a melt. The very small filaments are single crystals and very pure. Extremely high tensile strengths have been found in these whiskers. The improved physical strength of small wires has been utilized in the past by incorporating many fine wires into a cable. The latter is found to be much superior to a solid rod of the same diameter.

4. *The quality control principle.* With the industrial revolution came the use of machinery for the large scale production of goods. It became possible to make each item very similar to all those produced before. Not only were its size, shape, and performance nearly identical to its neighbors, but the amount of material, quality of material, input of labor, and even its color were as closely matched as possible. When production rates were low, it was possible to test and evaluate each unit. With increased throughputs, however, it became necessary to adopt statistical principles. The latter specify the degree of uniformity to be expected when samples are taken and used to judge the whole. If one out of every 1,000 units is sampled and found correct, statistical methods enable the manufacturer to determine with what degree of certainty he can assume that the entire batch of 1,000 is correct.

5. *The underrating principle.* When a part of a machine must not fail within some preset life expectancy, it is often possible to apply a "safety factor." The part might be made thicker than necessary for its present service; the excess material will provide additional protection against accidental overload or unforeseen stresses. The part is now underrated for the value of the stresses to which it is to be exposed. Another example would be an incandescent lamp which must last for a longer-than-normal period and is therefore run at reduced voltage. It produces less light but its filament evaporates at a greatly reduced rate. Lamp life is thereby extended by many hours.

6. *The mixing, alloying and adjoining principle.* If a material does not have the properties we desire, we can add other materials to modify it. In the simplest case, two or more materials are blended to form an ordinary mixture; the final property is linearly related to the composition of the mixture and the individual properties of the constituents. In other cases, the final property varies non-linearly with the amount of additives. It is sometimes possible with these

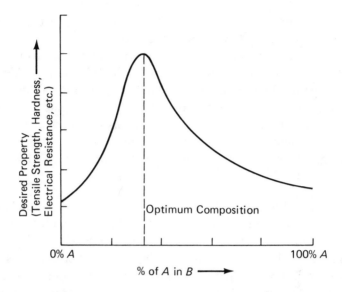

Fig. 2-5. Properties of some mixtures as influenced by composition.

mixtures to find an optimum composition for a particular purpose (Fig. 2-5). The materials, after mixing, may react chemically so that a new substance possessing the desired characteristics is formed. Alloys are metallic blends which comprise interlocked crystals formed during solidification. All the constituents of the alloy are mutually soluble when the mixture is molten, but not when it is cooled. The crystal structure of an alloy is sometimes very complex, consisting of pure metals, partial solutions of one metal in another ("solid solutions"), metals chemically reacted with one another, and entrapped impurities. After the alloy is formed into the desired shape, its composition is often further altered by various heat treatments. In some cases, the addition of a very small amount of one metal to another has a great effect on the latter's properties. Less than 1% of arsenic added to a melt of copper, for example, significantly increases the hardness of the copper. Small traces of rare earth metals added to silicon greatly alter the electrical properties of the latter. The change of material characteristics brought about by trace amounts of additives depends for its effectiveness on a technology which is able to produce ultrapure raw materials, and on very sensitive analytical methods by which to control the additions.

In some instances, it is unnecessary or undesirable to intimately mix two or more materials prior to fabrication; it is sufficient to laminate them. The materials are glued, welded, riveted, or strapped together in two or more layers. The assembly has some of the properties of its constituents but will also possess a number of superior characteristics. The designer of a laminate can achieve a wide range of properties in a composite by varying the components, their relative thicknesses, and their arrangement.

7. *The equilibrium principle.* It was stated above that all systems tend in the long run to revert to "maximum disorder" or chaos. Over the short term, however, a system will tend to remain in its existing state. Efforts to alter the latter will encounter resistance. The faster one tries to alter the existing state, the greater will be the resistance. Newton's classic first law of motion states that a body will tend to remain in its condition of rest or motion unless acted upon by an outside force. The object at rest or in uniform motion is in a state of equilibrium. To disturb this equilibrium, force must be applied on the object. The body will resist any attempt to destroy its equilibrium. A compressible object at rest will deflect under an external force before it starts to move. After uniform motion is reached, a new state of equilibrium is established. To change speed (acceleration or deceleration) again involves a disturbance of equilibrium and the application of additional force. Non-mechanical systems in equilibrium also display this property of resistance to change. A study of the equilibrium of a particular system will often permit the arranging of conditions to achieve an optimum result. Assume, for example, that we are dealing with a continuous chemical process in which one of the products is a gas. The reaction is being carried out in a closed tank. It will reach an equilibrium which depends on the rate of gas removal. To increase the rate of the reaction, the gaseous product is pumped out of the tank at a faster rate. If the reaction is an undesirable one and we wish to slow it down, the pressure on the system is increased by the addition of an inert gas; this tends to inhibit the reaction. Similarly, if a process gives off heat, it can be accelerated by cooling and decelerated by heating.

8. *The saturation principle.* Another way to slow or prevent the occurrence of an undesirable reaction is to place some barrier in the path of the participants. To prevent the rusting of steel, e.g., the most direct method is to cover it with an adhering substance which will itself resist the attack of atmospheric oxygen. A more subtle approach is to purposely expose the bare metal to an oxygen-rich chemical. The resulting high rate of rust formation on the metal surface produces a thin layer of tightly interlocked rust crystals which now act as a shield against further attack. We have saturated the surface so that it is no longer chemically active. By the sacrifice of a very small amount of the material to oxidation, the remainder is protected from further attack. The saturation principle automatically comes into play when plastics are manufactured. The chemical activity of the raw liquids from which plastics are produced is used to agglomerate individual molecules into long chains made up of many thousands of molecules. The formation, tangling, and interlocking of these chains are, in fact, what causes the liquids to congeal into the solid plastic material. The chemical attraction of each molecule for outside substances is satisfied by its seizure of an adjoining molecule. As a result, the entire body of the plastic is chemically saturated. Because of this saturation, plastics will resist attack by many substances — atmospheric oxygen, water, soil constituents, etc. Plastics, when coated onto other materials, make excellent protective covers.

9. *The prestressing principle.* In some applications, a structural member is required to maintain a static or dynamic load for long periods of time. A bridge girder, for example, must hold the dead weight of the structure indefinitely. High pressure gas containers must retain their tensile strengths over long storage periods. The obvious response of a designer to the requirement for more load carrying ability is to provide greater thicknesses or to add reinforcing members at critical points (*the underrating principle*). These pro-visions are sometimes of limited usefulness, however, because the weight of the additional material increases the overall load the member must carry. In addition, long term loading produces metal "fatigue" and creep, both of which reduce the ultimate strength of the load supporting member. A satisfactory solution to the problem is to artificially pre-load the member so that is has a permanent stress in a direction opposite to the

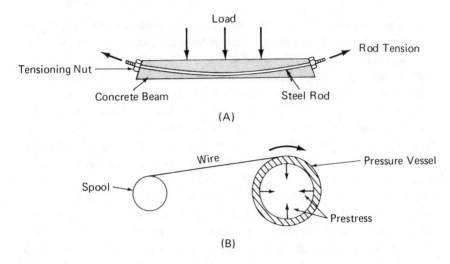

Fig. 2-6. The prestress principle.

one to be encountered in the application. Fig. 2-6 indicates two methods for utilizing the prestressing principle. As applied to structural beams, the method consists of casting a bent rod of high tensile strength steel into the concrete of the beam. The rod is threaded at both ends. Before the concrete has cured, nuts on the ends of the rods are tightened to slightly straighten the rod. This produces upwardly directed stresses (which will be permanently retained) in the beam. When the beam is placed in service, the load-induced stresses tend to cancel the prestress. The member can now serve its purpose indefinitely with no fatigue. As applied to pressure vessels, the method consists in mounting the vessel in a lathe and winding it with wire or metal tape under high tension. This produces "hoop" stresses directed inwardly. The stress produced when the vessel is filled with gas at high pressure cancels the prestress, and the container metal is essentially at zero stress. The prestressing principle permits the use of lighter construction members and pressure vessels having thinner walls. Fatigue and creep are eliminated or considerably reduced.

10. *The self-generating principle.* Some times unusually high accuracy or efficiency can be obtained by clever utilization of system characteristics. At other times, a system can be so coupled as to aid in its own operation. The self-generating principle may be defined as the art of utilizing part of the forces already present in a device for

its overall guidance or for an improvement in its performance. It is a kind of feedback without the use of descrete amplifying elements. Several examples will make the principle clear. If a number of steel cubes are enclosed in a rotating barrel along with water and an abrasive powder, there will be a wearing down of the edges of each cube. The edges and corners will be removed first because they are of the smallest area, and impact pressure there during each cube-to-cube contact is highest. Eventually, the cubes will reach spherical shape and begin to decrease uniformly in size; it will be found that each sphere is highly uniform in terms of its own diameter (although there will be considerable variation from sphere to sphere). It is, however, only necessary to separate the units by size in order to form a stockpile of precision ground balls (useful for bearings and measuring standards). Production can be considerably speeded by starting with spheres which have been formed by the rough pregrinding of steel cubes. The tumbling action in the barrel creates a condition whereby each ball serves as the grinding tool for its neighbors. The grinding process is also the "precision-inducing" process. The production of precision spheres by other means would require much more elaborate machinery than a tumbling barrel.

Another example of the self-generating principle is the use of a soft seal to contain high pressures. One form is shown in Fig. 2-7: the upper plug is used to close off a high pressure vessel but must be removed frequently for emptying and recharging. To make the system self sealing a slotted flexible ring is positioned around the plug before the latter is inserted and bolted down. When pressure is applied, gas enters the slotted portion of the seal ring and causes radial expansion against the walls of the vessel; the higher the pressure, the tighter the seal. Because the pressure itself does the work of sealing, it is unnecessary to apply more than a minimal torque to the hold-down bolts.

A final example will illustrate an entirely different application of the self-generating principle. A difficult military problem before the days of air transport was the moving of men and equipment across wide and fast flowing streams. An ancient but elegant solution makes use of the river's force; the method is illustrated in Fig. 2-8. The first step is getting a strong swimmer across the stream with one end of a light cord. When this has been done, several strong "pendulum" lines are brought across and anchored firmly to a large tree or rock at point A. A raft is then constructed and lashed to the

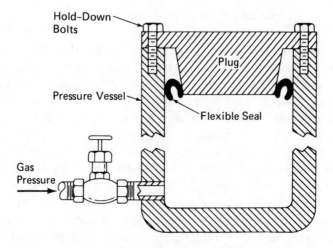

Fig. 2-7. The self-activating seal.

pendulum lines. A control line is tied to the raft and snubbed around a tree at point *B*; the raft is now loaded. The control line is carefully paid out; the force of the flow will move the raft and its load across the river. When unloaded, the empty raft is easily returned for another load by use of the control line. Employing a long

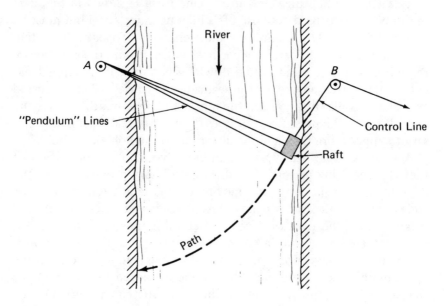

Fig. 2-8. The "pendulum" method for harnessing river power.

control line, the last man (or team) can use the raft to cross the river. All lines and the raft are then recovered and stored for later use.

The Associated Art Effect

Many significant and operationally valid inventions have failed to sell because they "were ahead of their time." In practical terms this phrase often meant that suitable construction materials were not available or systems that could use the invention were not in existence. In some cases systems in which the device could be used were being developed, but this was unknown to the inventor.

If the associated art is studied at the same time that the invention first shows signs of feasibility, it is often possible to make additional inventions or to change existing conditions to accommodate the invention. The gasoline turbine, for example, lay idle for a long time because the steels having the desirable low-creep characteristics at high temperatures were not yet developed. When these steels did become available, major manufacturers of internal combustion engines were tooled up for or otherwise committed to the piston type. If creep studies had been conducted earlier, it might have been possible to utilize the relatively simple gasoline turbine in early automobiles.

Thomas Edison was farsighted enough to realize the associated art requirements of his electric lamp. He designed generating and distributing equipment to make up a complete system. His demonstration plant in New York provided lighting for several square blocks and set a pattern for modern lighting systems.

The physicist Oliver Heaviside realized that his work on ionization layers in the upper atmosphere could not be made truly quantitative because no known form of mathematics would explain the phenomena with which he was working. Not to be held back by this lack of associated art, Heaviside invented his own calculus, which he then used to develop important principles in long-distance radio communication.

More recently, the inventors of a new ultraviolet lamp could not find a market for their improvement even though its output for a given size was much greater than existing types. Fortunately they discovered a codeveloping art: ultravoilet drying inks. The latter showed great promise for printing on metal surfaces such as panels and drink containers. The joining of the two inventions then led to rapid acceptance of the new lamp. More details on this lamp will be found in Chapter 11.

3

The Anatomy
of a Patent

The Patent

In this chapter, we will examine the structure of patents, their properties, and some of their uses.

A patent is a contract between the government and the inventor, granting the inventor, for a fixed period, the sole right to the "enjoyment" of his creation. A patent sets up a monopoly; the holder can legally restrain anyone from making, using, or selling the invention.

The need for a patent system was considered so important by the founders of the United States that provisions were included in the Constitution. In Article I, Section 8, we find:

> The Congress shall have power to promote the progress of science and useful arts by securing for limited times to authors and inventors the exclusive right to their respective writings and discoveries.

At the second session of the first Congress, an act was passed authorizing a patent to be issued to the inventor of any useful art, etc., on his petition, "granting to such petitioner, his heirs, administrators or assigns, for any term not exceeding 17 years, the sole and exclusive right and liberty of making, using and rendering to others to be used, the said invention or discovery."

The issuance of a patent represents the transition of an idea from an ephemeral concept to the status of a legal entity. For a period of

17 years, an issued patent is true property. It can be sold, leased, or transferred to another. If the owner dies, it becomes part of his estate and is inheritable. If a profit is made in the sale of a patent, the profit is taxable in the same way that profit from the sale of merchandise or real estate would be taxable.

The right to patent is open to all citizens. When and if certain requirements are met, a patent is issued to the inventor; he is then free to use the patent in any way that he sees fit, subject only to the restrictions of maintaining the general public well being.

At the present time, there have been over four million patents granted by the U.S. Patent Office. There are approximately 100,000 applications per year, resulting in about 65,000 patents. Graphs of patent activity reflect economic conditions; that is to say, they follow the cyclic trends of business. Notwithstanding some decreases during business recessions, the overall trend in the number of patents granted each year is upwards and is a measure of the increasingly complex technology associated with our present day lives.

Structure of the Patent

The independent inventor will most likely retain an attorney or patent agent to handle the writing and legal parts of his application, but it will be desirable for him to develop skills in reading and understanding the patent literature on his own. The number of uses to which the latter can be put makes a knowledge of the format extremely valuable.

A patent consists of two major parts: (1) a certificate which actually grants the right to exclusive enjoyment of the invention and (2) a sheaf of documents which describe the invention. The certificate is of interest only to the owner or purchaser of the patent. The descriptive papers, on the other hand, are what is normally referred to as the patent. Prints of these papers are on file in Washington, D.C. and in many libraries.

If we examine a patent description carefully, we find that it is made up of nine components:

1. *The title structure.*
   a. Title of the invention.

b. Name and address of the inventor or inventors
c. Issuance number of the patent
d. Date of issuance
e. Date when application was made
f. Serial number assigned during the time when the application was in process (this number was not available to the public until the issuance date).
g. If the patent is assigned during any time prior to its issuance (i.e., rights transferred to another), the name of the assignee
h. Class and subclass of the invention and the field of search. These numbers refer to a classification manual which contains over 300 general titles by which inventions can be described. Classification will be discussed in a later section.
i. References. A partial list of patents which appear to have some prior-art relationship to the patent in question. References may be of historical importance, reveal inventions which are similar to but not identical with the present one or they may represent the application of similar principles in some other field.
j. The name of the patent examiner.
k. The name of the attorney or patent agent.
l. The number of claims and drawing figures.

2. *Abstract.* A concise summary of the invention. The abstract describes the invention but does not indicate its advantages or applications. It is usually 100 words or less and one paragraph in length. The abstract appears at the beginning of the patent when it is published. The abstract and one drawing are also reproduced in the Official Gazette of the U.S. Patent Office. The gazette is issued weekly; back copies are kept in libraries of major cities in the United States. The abstracts permit rapid, preliminary evaluation of individual patents.

3. *Background of the invention.* A description of the specific area in industry, agriculture, etc., to which the invention applies. A discussion of why the invention was made, the need to be met, or the improvement which has been lacking.

4. *References to related inventions, if any.* How presently-patented devices solve the problem but still leave something to be desired.

5. *Summary of the invention.* A statement of the means by which the problems outlined under "background of invention" have been solved. The reasons for adopting this particular means are given as well as the advantages and further applications of the invention.

6. *A brief description of the drawings.* Each drawing is identified by a number and a general statement as to what it is — e.g., "a front view of modification 1;" "a cross section along the line *A-A* of the previous figure;" etc.

7. *Detailed description.* This is a statement of the construction and operation of the invention. The aim is to teach others in sufficient detail so that they may understand, be able to construct, and make use of the inventor's concept. This section also reveals the preferred embodiments of the invention. The inventor must choose these carefully; afterthoughts and later improvements are not admissable once the application has been sent into the Patent Office. The description will often identify the best mode of carrying out the invention from among the preferred embodiments.

8. *Claims.* These are a series of numbered statements spelling out what the inventor feels (and the Patent Office has allowed) to be the novel features of his invention. The claims must represent logical extensions of what has been revealed in the detailed description. If a claim is made relating to an embodiment which has not been included in the detailed description, it will be rejected by the patent examiner. If, on the other hand, some variation of the invention is included in the description but no corresponding claim has been made, the patent will issue with fewer claims than the inventor is entitled to have. If this situation is not corrected by a "reissue" patent (granted later), the inventor will have "donated" that part of his invention to public use and can no longer protect it. The writing of effective claims requires a high degree of skill, which can only be achieved through experience.

9. *Drawings.* The Patent Office encourages the use of drawings whenever possible. These may be shaded isometric representations, cross sectional views, or plane projections. In chemical patents, formula drawings representing compounds are used. Circuit and logic diagrams are acceptable in certain types of electrical inventions.

The relatively simple patent of Figs. 3-1A, B, C, and D will be used to help the reader identify the above listed elements in an actual case. The invention relates to an attachment for a golf club which will enable the user to recover a ball from an area of difficult access. The patent number is 3,749,407 and the date of issue is July 31, 1973. The inventor's name, Prochnow, and his address are given but no assignee is recorded. In order for an assignee to be shown on the patent, notice of assignment must be sent to the Patent Office prior to the issuance and publication. Transfer of the rights to a patent can be accomplished at a later date but must be registered with the Patent Office to be valid against a subsequent assignee. The original application for this patent was filed on April 12, 1971, so that slightly more than two years and three months were required for the entire process. The heretofore confidential application number was 133,182. In all communications with the Patent Office between the time of filing and the notification of issuance, the inventor's attorney or agent used the latter number to identify this particular application.

The classes searched for prior art were:

a.   Class 273 subclasses 162E, 32F.
b.   Class 294 subclass 19A and 19R.

The references cited show a list of patents, their dates of issue, inventors, and classification numbers. These patents represent "prior art." This list is not to be considered complete or all-inclusive; it shows what the examiner deemed as applying most closely to the application in question. The reference patents make valuable reading for anyone who wishes to conduct further work in this area or wishes to trace the line of thought of successive inventors.

The examiners and the attorneys for the inventor are identified next, followed by the abstract and a specimen drawing. It is this

United States Patent [19]

Prochnow

[11] **3,749,407**

[45] **July 31, 1973**

[54] **BALL RETRIEVER ATTACHMENT**

[76] Inventor: **Lee W. Prochnow,** 4103 W. 98th St., Chicago, Ill. 60453

[22] Filed: **Apr. 12, 1971**

[21] Appl. No.: **133,182**

[52] **U.S. Cl.** .... 273/162 E, 294/19 A, 273/DIG. 4
[51] **Int. Cl.** A63b 57/00
[58] **Field of Search** .... 273/32 F, 162 E; 294/19 R, 19 A

[56] **References Cited**

UNITED STATES PATENTS

2,432,906	12/1947	Klingler	294/19 A
3,058,767	10/1962	Baker	294/19 A
2,561,815	7/1951	Oberg	294/19 A
3,210,111	10/1965	Fallon	273/162 E X
D216,138	11/1969	Carignan	273/162 E X
1,431,968	10/1922	McDermott	294/19 A
1,674,294	6/1928	O'Rourke	273/162 E X
2,448,644	9/1948	Williams	294/19 A
2,523,942	9/1950	Ciambriello	294/19 A

Primary Examiner—Richard C. Pinkham
Assistant Examiner—Richard J. Apley
Attorney—Hill, Sherman, Meroni, Gross & Simpson

[57] **ABSTRACT**

This invention relates to a cup-like device having drain openings therein which is made as a unitary structure by injection molding of a flexible elastic plastic such as low-density polyethylene, polyvinyl chloride (PVC), or the like. The cup may readily be secured to, or removed from, a golf club head without the use of tools and when so secured may be used to retrieve golf balls from relatively inaccessible locations or to pick up golf balls after a practice session. Attachment and removal of the cup from the golf club is accomplished by insertion or removal of the club shaft by deflecting a flexible portion of a clip member integral with the cup and then moving the clip and cup along the shaft into engagement with or detachment from the golf club head.

2 Claims, 4 Drawing Figures

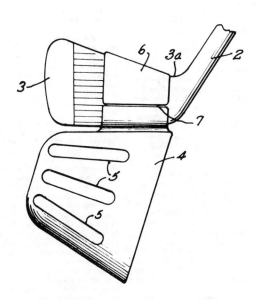

Fig. 3-1A. Patent for a ball retriever attachment.
Title structure, abstract, sample drawing.

1

BALL RETRIEVER ATTACHMENT

BACKGROUND OF THE INVENTION

SUMMARY OF THE INVENTION

The desirability of providing a cup-like retriever with an elongated handle for retrieving golf balls has long been recognized. Such devices are particularly useful for retrieving golf balls from water hazards, areas of ground under repair, muddy areas, and other places difficult of access. They are also useful in picking up previously hit balls from a practice area.

While such devices previously have been provided for attachment to golf club heads, some have been objectionable in that they require operation of a thumb nut for attachment to the club head and such attaching means frequently loosen in use. In those integral retrievers previously provided, the cups have been inadequate for the required purpose or the attachment to the club has been insecure.

The present invention overcomes the disadvantages of the prior art by providing a drainable cup of ample size molded of an elastic plastic which has an integral clip member formed therewith that may be engaged between the angle between the hosel of the club and the inner end of the club head and an outer portion of the club head where it and the clip associated therewith are firmly held in position for use.

The retriever is attached to the club by insertion of a club shaft at a point above the hosel through a slot in the side of the clip, the elastic clip material being distorted during such insertion to permit the shaft to enter the clip and then returning to its original position to retain the shaft within the clip. The clip and its associated cup are then pushed downwardly over the hosel and outwardly until the clip completely engages the club head between a portion inwardly of the outer end of said club head and the curved portion where the head joins the hosel. In such position, the clip, which is formed to fit the club head, is held firmly in place on the head and the cup formed therewith firmly held in position for use. Due to the clip being held between the angle at the juncture of the hosel and the club head and the outer end of the club head, which increases in size outwardly from such angle, dislodgment of the retriever in normal use is prevented. At the same time, the retriever may be detached from the club by a reversal of the procedure above described. In removal, the clip is forced outwardly and upwardly over the angle up over the hosel and around the shaft where it may be removed by depressing the side of the clip at the slot and moving the shaft through the opening thus provided, after which the elastic material of the clip causes the depressed side portion thereof to return to its original position. Since the clip and cup are formed of flexible elastic material they may be compressed for storage in the golf bag or practice ball bag, as desired.

It is an important object, therefore, of the present invention to provide a golf ball retriever which readily may be attached to, or detached from, a golf club without the use of tools, which is inexpensive to construct, and which remains in firm association with the golf club to which it is attached for use.

Other and further important objects of this invention will be apparent from the following description and the accompanying drawings.

2

BRIEF DESCRIPTION OF THE DRAWINGS

The invention (in a preferred form) is shown on the drawings wherein:

FIG. 1 is a side elevation showing a golf club head with the improved ball retriever of this invention attached thereto;

FIG. 2 is a perspective view looking at the structure of FIG. 1 from the left in FIG. 1;

FIG. 3 is a side elevation of a portion of a golf club and the improved ball retriever cup of this invention in an initial position of installation; and

FIG. 4 is a view along the line IV—IV of FIG. 3.

DETAILED DESCRIPTION OF THE DRAWINGS

As shown on the drawings, the reference numeral 1 designates the shaft of a golf club to which is connected a hosel 2 that, in turn, carries a head or blade 3. Due to the construction of the so-called golf "iron" the angle of juncture between the hosel 2 and the blade 3 provides a protuberance generally at 3a. The retriever of this invention is designed for use with so-called "long irons" such as No. 1, No. 2 or No. 3 and since except for loft these irons have blades or heads 3 of substantially the same shape the attachment of the retriever thereto may be effectively and interchangeably accomplished with any of these clubs. This selection of the so-called "long irons" is in part due to this similarity of the head shape and also due to the fact that such "long irons" have longer handles and thus increase the distance at which the retriever may be operated.

The retriever attachment of this invention consists of a unitary structure which is injection molded of a flexible elastic plastic such as low-density polyethylene, polyvinyl chloride (PVC), or similar plastics which are well known in the art and which are capable of being molded but which when molded are relatively flexible and elastic.

The retriever cup of this invention is designated generally by the reference numeral 4. Drainage from the cup is provided by a plurality of slots 5 and also by an aperture at the bottom of the cup (not shown) which is of relatively large diameter but of a diameter smaller than the largest cross-sectional area of a golf ball to be retrieved.

Formed integrally with the cup 4 is a member 6 which I have designated generally as a clip member. This clip member 6 is provided with top and side walls and the juncture thereof with the cup provides a bottom wall, so that the clip in effect provides an open end enclosure, the area of which increases toward its open end so that it may interfit with the blade or head 3 of the golf club. One side of the clip is provided with a slot 7 which permits of the wall of the clip to be depressed above the slot by pressing thereon at a point 7a. This depression of the side wall as shown in dotted lines in FIG. 4 permits of insertion of the shaft 1 within the confines of the clip, after which the depressed portion snaps back into the position shown in full lines in FIG. 4 so that the clip encloses the shaft and prevents the retriever being lost in the event of improper attachment thereof to the club head.

In installing the retriever for use, the clip 6 is first depressed to permit insertion of the shaft 1 as shown in FIG. 4, at which time the parts are generally in the position shown in FIG. 3. Thereafter, the clip and its integral retriever cup 4 are moved downwardly along the

Fig. 3-1B. Patent for a ball retriever attachment.
First two columns of text.

3

shaft, over the hosel 2 and into the position shown in FIG. 1. The angularity of the clip 6 permits the same to fit against the flare of the club head or blade 3 and the clip is then held firmly between its outer end which is in engagement with the club head or blade 3 and the protuberance at the angle 3a above described. Due to the flexibility and elasticity of the material from which the clip is molded, the same in the position shown in FIG. 1 provides a tight fit with the club head or blade 3 so that it is difficult to dislodge the retriever in use.

To remove the retriever from the golf club a reversal of the procedure above described is followed. The clip is forced outwardly and upwardly over the angle protuberance 3a and the hosel 2 until it again encircles the shaft as shown in FIG. 3. At this point the top portion of the clip 6 above the slot 7 is depressed by pressing on the clip approximately at the area shown at 7a so that the clip again assumes the dotted line position shown in FIG. 4, the shaft may be moved out of the clip, and the retriever is stored for subsequent use. Due to its flexibility and elasticity, the retriever may be readily compressed for packing within the pocket of a golf club bag or other container such as a golf ball bag used by the golfer.

As will be apparent from the foregoing description, this invention provides a convenient retriever device which may be carried in a golf club bag, ball bag, or the like and readily installed on a "long iron" for use when it is desired to retrieve a golf ball from an inaccessible area, such as a water hazard, ground under repair, muddy water, poison ivy, thick vegetation, bushes and thorny thickets where the golfer either cannot reach the ball or would be injured or poisoned in so doing.

The retriever of this invention is inexpensive, cannot become lost even if improperly installed, is provided with ample drain area for the draining of water therefrom, and convenient to use in the picking up of golf balls under the foregoing circumstances or after a practice session, eliminating the necessity of the laborious effort of bending over to pick up the balls.

I am aware that details of construction may be varied and I, therefore, do not purpose limiting the patent hereon granted other than as indicated by the scope of the appended claims.

I claim:

1. In combination, a golf ball retriever and a golf club having a shaft, a head on the free end portion of said shaft having a hosel fitting about the free end portion of said shaft, and a tapered metal blade extending from said hosel with a front striking face and a rear face, having a sole adapted to lie parallel to the ground, a top surface tapering outwardly with respect to the sole as it extends from the hosel, and a blade tapering to con-

4

verge from the sole to the top surface thereof,

a foldable plastic cup molded of flexible plastic material having a flat bottom of the general diameter of a golf ball and an annular side wall tapering upwardly from said bottom and terminating into an opening of a size to accommodate the ready scooping of a golf ball from water and the like,

said side wall having a plurality of slots extending therealong for the drainage of water from the cup, and having an integral clip extending radially outwardly of said side wall at the angle of taper thereof,

said clip being in the form of an open loop having a bottom surface extending along the outside of said side wall and having generally plane surfaces converging relative to said bottom surface from opposite sides of said bottom surface as they extend outwardly therefrom and having a rounded outer end portion tapering outwardly from the upper end of said clip to the lower end thereof to correspondingly mate with said top surface of the blade of the golf club, and with said bottom surface engaging said sole of the club head, said open loop being formed by one of said converging plane surfaces, said one converging plane surface having a top flange portion extending from said rounded outer end portion, a slot extending along its entire surface, and a bottom inturned flange portion extending from said bottom surface, wherein said top and bottom flange portions engage one of said faces of said blade,

said inturned portion forming a face adapted to be engaged by the shaft of a golf club and accommodating said clip to be depressed by pressing against the shaft of a golf club and fit said clip over the shaft to be slidably moved along said shaft and hosel to fit along the blade of the golf club and support said cup to be positively retained to depend from the blade of the golf club in relatively rigid relation with respect thereto by the form of said clip and the elasticity of the material from which said cup is made, to position said cup to scoop up a ball by movement of the club head with the hosel leading the blade of the club head, the foldability of said cup enabling said cup to be readily carried in the pocket of a golf bag.

2. The golf ball retriever of claim 1, wherein the plastic material is a low density polyethylene of sufficient elasticity and flexibility to snap into firm engagement with the blade of the club head and still enable the retriever to be folded to a compact size.

* * * * *

Fig. 3-1C. Patent for a ball retriever attachment.
Last two columns of text, including claims.

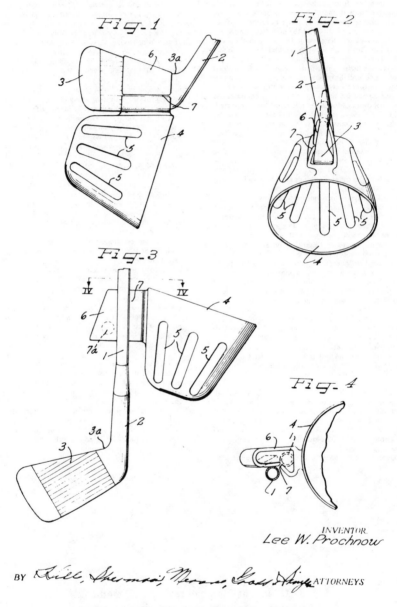

Fig-1

Fig-2

Fig-3

Fig-4

INVENTOR.
Lee W. Prochnow

BY *Hill, Sherman, Meroni, Gross & Simpson* ATTORNEYS

Fig. 3-1D. Patent for a ball retriever attachment.
Sheet of drawings.

statement and drawing which are published in the Official Gazette of the U.S. Patent Office. A searcher does not need to read the whole patent document or look through all the drawings to obtain a general idea of the patent's content.

In Fig. 3-1B is shown the first sheet of text of the full patent. Each sheet contains two columns and the lines are numbered for easy referral. The background of this invention is presented in column 1, lines 5 through 20. The need for such a device is noted, as are some previous solutions to the problem. The difficulties in using these previous inventions are pointed out (e.g., the need for a thumb nut and its loosening in use). In lines 21 to 58 of column 1, the invention is summarized. A word picture is presented showing how the invention is constructed (molded of an elastic plastic with an integral clip member) and how it is attached, detached, and stored. A preliminary discussion of the objectives of the invention is also given in column 1 (lines 59 to 67) to be followed by a more thorough description later. In column 2 of the patent, lines 3 to 13, is a brief description of the drawings. This tells the reader only what each drawing represents in general, and acts as an inventory so that no drawing will be overlooked. The drawings themselves are shown in Fig. 3-1D; it is easy to check them against the brief description in column 2 to see that all drawings are accounted for and that the reader understands all the views. Figure 2 of the patent can be identified as a perspective of the retriever, as shown in plan view in Fig. 1. Figure 4 of the patent is what would be seen by looking down along the section IV-IV of Fig. 3. Several characteristics of patent drawings can be pointed out from these examples. Shading is used whenever possible, with the source of light considered to be coming from the top left corner at an angle of 45°. Reference to individual parts is by number; like numbers in different figures refer to like parts. For example, the number 3 referring to the club blade in Fig. 1 of the patent must always be used for the blade in any of the other figures.

In the detailed description of the drawings (column 2, lines 15 to 68; column 3, lines 1 to 24), the various views are carefully explained with full use made of the individual numbers. In the process of describing the drawings, the writer of the patent adds many other details which teach others to make, use, and benefit from the invention. In column 2, lines 32 to 38, for example, the preferred method of construction and the preferred materials are mentioned — injection molding of low density polyethylene or polyvinyl chloride. More details on the functioning and advantages of the invention can now be presented because of the use of the drawings. In column 2, lines 40 to 46, and in Figs. 1 to 3, it can be seen that the cup is provided with slots and a perforated bottom to permit drainage. A special advantage of the construction described is that provided by clip 6 (column 2, lines 46 to 62, patent Figs. 2 and 4). This constructional feature permits the shaft 1 to be readily inserted into the retriever and also prevents loss in case of improper attachment. In column 3, lines 2 to 10, the clip 6 is described as having "angularity" — its taper matches that of the club blade 3 so that it is difficult to dislodge the retriever when it is used to scoop up a ball. The description continues in column 3, lines 25 to 40, pointing out some advantages to be gained by use of the invention. Aside from being able to save a ball which would otherwise be inaccessible, the device prevents the golfer from possibly being endangered by contact with poison ivy. The invention is also labor saving during practice sessions (presumably allowing more of the user's energy to be applied to acquiring proficiency in the sport). The section ends with a broad statement (column 3, lines 41 to 44) which says, in effect, that the detailed description is somewhat generic, that some details might be varied without departing from the general idea of the invention (thus not circumventing the patent), and that the claims only shall be considered as the limitations on the scope of the disclosure.

It may occur to the reader studying patents for the first time to ask why the writer of a patent should go to such lengths to teach others what has been a relatively secret matter up to the point of filing the application. The tutorial aspects of the patent would appear to make it easier for a would-be circumventer to discover points of weakness. The logic of a thorough disclosure is based on several factors. In order for a patent to be granted on an invention, the applicant must show that the improvement was not made by mere exercise of technical skill and that the improvement would not have been obvious to anyone skilled in the art. The improvement must not represent merely the use of superior materials, simple change in form or arrangement of parts, or the addition of a power drive to what was formerly a hand operation. Additionally, it cannot be a mere aggregation of parts. If the improvement for which the inventor seeks a patent is sufficiently subtle so that it avoids the above objections, then it is a true invention, but the applicant must convince the examiner that he has overcome these limitations. He must carefully instruct all others in its construction, functioning, and advantages. In this way, the inventor's contribution is carefully delineated from the work of the skilled artisan. By protecting his results in the form of a patent, the inventor stakes out a definite area which is then guaranteed by law.

The claims are without doubt the most critical part of the patent. Like claims to land acreage, these statements define exactly what the inventor has succeeded in fencing off as his property. On inspecting claim 1 (Fig. 3-1C, column 3, lines 45 to 53; column 4, lines 1 to 7) we see that it is a single sentence! This structure of one sentence per claim is a legal requirement. The fees charged by the Patent Office depend on the number of claims. It thus benefits the inventor, at least initially, to minimize the number of claims and to lengthen each one as much as needed to fully cover the improvement. In a lengthy sentence describing some feature, it will often be found that the same item is mentioned several times. It is common practice to use the word "said" prior to an item when it has been introduced earlier in the sentence. Thus, in column 3, line 47, the golf club's shaft is first mentioned; in line 48, it is mentioned again on the next line but now preceded by the word "said." This occurs again in line 49. This practice is only continued where necessary to prevent vagueness. In other parts of the claim (e.g., column 4, line 33), the word "shaft" appears alone. There is no need here to refer to its previous usage. Once the beginner notes this peculiarity of claim sentence construction, he will find it much easier to read and comprehend patent claims.

After describing the golf club which is to be used in combination with the invention, claim 1 details the invention in neat and logical order. It lists a cup which is both flexible and foldable (column 4, line 2). The cup has sides tapering up from its bottom and the sides are slotted to allow drainage of water. The top of the cup is of such a diameter as to permit scooping up a golf ball (column 4, lines 5 to 7). From the sides, there extends a loop (somewhat similar to the handle of a drinking cup, except that the area enclosed by the loop extends horizontally rather than vertically). The loop is slotted (7 in Fig. 4 of the patent and tapered so that it fits the tapered blade of the golf club; the loop and slot form a clip, which permits the retriever to be pressed against the shaft (column 4, lines 13 to 37). When the flexible slot gives way to accommodate the shaft within the loop, the retriever can be slid along the shaft to be firmly positioned along the blade of the club (column 4, lines 36 to 40). The mode of operation is also indicated in this claim wherein it is stated that the cup scoops up a ball when the club is moved so that the hosel (the cylindrical part of the blade casting) leads the blade (column 4, lines 43 to 45). The advantage of using an elastic material is mentioned in lines 45 to 47 of column 4; the retriever can be folded up and stored in a golf bag.

Claim 2 is of a type called a dependent claim. It doesn't extend the inventor's rights by any other embodiment or modification; it merely expands on a previously claimed feature — the use of a specific flexible and foldable plastic. In this case, the material mentioned is low density polyethylene.

To go just one step further in studying the above patent, let us look at one of the references to prior art which it lists. In Fig. 3-2 appears the sheet of drawings for Patent Number 2,523,942 — the last reference found in Fig. 3-1A. This patent, issued to Ciambriello, lapsed in 1967 and was in the public domain at the time Prochnow's application was filed. The Ciambriello device differs in that it is used with a forward, prodding motion, rather than as a scoop, and it fits onto the handle of a golf club, rather than on its blade. The golf ball is retained by indentations, 16, after it has been forced into the funnel-like and flexible structure, 11. The open structure of the latter leaves no question about water drainage. A ball trapped in the retriever is removed by finger pressure applied through the opening, 18 (Fig. 2 of the patent). The device as shown requires a screwdriver for attachment. It cannot be folded but must be stored as is. These two characteristics are disadvantages which have been identified and overcome by the Prochnow patent. An interesting feature of Ciambriello's retriever is shown in Fig. 5 of the patent. Here the invention is shown attached not to a golf club but to a telescoping structure of its own. The telescoping rod, 22, can be extended for reaching more remote locations or retracted for storage along with

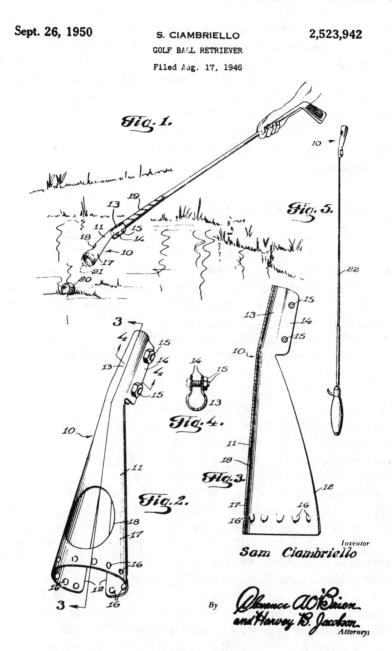

Sept. 26, 1950

S. CIAMBRIELLO

2,523,942

GOLF BALL RETRIEVER

Filed Aug. 17, 1946

Fig. 1.

Fig. 5.

Fig. 4.

Fig. 2.

Fig. 3.

Inventor

Sam Ciambriello

By Clarence A. O'Brien
and Harvey B. Jacobson

Attorneys

Fig. 3-2. Sheet of drawings for a golf ball retriever.

the clubs. In this version, the retriever would offer a distinct advantage not obtainable with the later development by Prochnow.

It would be advantageous for any inventor to become skilled in the reading and interpretation of patents. Files of patents and the Official Gazette of the U.S. Patent Office can be found in the central libraries of most large cities; a listing of these libraries is given at the end of this chapter. Continued practice in reading will build speed and will facilitate the efficient recovery of information from this very large source.

Patents As Technical Literature

Ordinarily, patents are read to discover prior art and to determine whether a new improvement is sufficiently different to justify the expense of protecting it. Since the searching for prior art is usually done by a professional, the inventor is often unaware of the tremendous body of knowledge available to him and the various ways in which it can be employed. If the inventor were to do his own search, he would soon find a second way to utilize the patent literature. As devices similar to his were uncovered, he would find himself modifying his own concepts to veer away from territory established by others; he would also begin to improve his invention in terms of his growing knowledge. As he continued his search, he would find it very tempting to stray into other areas as new ideas occurred to him. He would discover that the patent literature resembles a huge variety store with many departments; it is difficult to maintain a single goal amid so varied an array of interesting merchandise.

The reader of patents soon acquires a tremendous "sense of knowledge" in the area in which he is working. This comes about as a result of the inherent nature of the patent. As was mentioned in previous sections, the process, machine, method of manufacture, or composition of matter described in a patent is not a minor improvement which could have occurred to anyone skilled in the art. It represents a true and unique advance in its particular field. Another requisite for a patent is that it be workable. Although the Patent Office no longer requires a functioning model for most applications, it is assumed that the applicant has physically tested the invention and found it to be functional. In a court contest concerning the validity of a patent or any of its claims, a demonstrated lack of workability would be deadly evidence against the inventor. For these three reasons — uniqueness, a step forward in a particular technology, and workability — the great majority of patents represent an excellent source of practical information.

We are constantly aware of the growing number of books and magazines published each year. Despite this vast number of publications, it is often not possible to obtain detailed information on some very small but important segment of technology. For example, books and articles describing carburetors are available. If we wished, however, to explore the design of components of outboard engine carburetors — needles, floats, or mixing chambers, we could find relatively few texts which even mention these subjects. The patent literature, on the other hand, is rich in detailed information on each of these components. The patents not only show how these items work but teach in a practical and painstaking way how they can be designed, produced, adjusted, and modified.

A further use of the patent literature is in tracing the progress of a particular development. The purpose may be the preparation of a historical text, the study of the mental processes associated with inventing, or the extrapolation of some recent development to some new level. In recent years, historical studies based on patents have appeared on such specialized topics as tunnels, air guns, subways, prosthetic devices, turbine-driven ships, and writing implements. These books are used by schools, collectors, and hobbyists, as well as by inventors.

The reasoning processes of inventive individuals have fascinated psychologists for many years. The results of these studies have been used to formulate problem-solving systems. Several of the most interesting of the latter will be discussed in a later chapter. To illustrate how well the creative process can be traced by means of patents, let us examine one highly successful invention — the tape recorder. In 1890, Valdemar Poulsen, a Swedish citizen, invented an apparatus for recording sound on a wire. His earlier models made use of magnetizable wire wrapped around a drum. This version resembled Edison's phonograph except that wire was substituted for the grooves made by the recording needle. Because of spacing problems, recording time with this arrangement was much less than that offered by the fine grooves of the Edison apparatus. Evolution was slow because the phonograph disc was

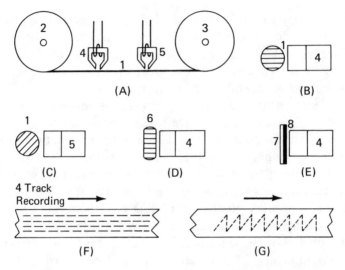

Fig. 3-3   Stages in the development of the tape recorder.

evolving rapidly and satisfactorily. By 1920, however, the wire recorder had been developed into the form shown in Fig. 3-3A. A steel wire, 1, traveled between two drums, 2 and 3. An "input" electromagnet, 4, was arranged to contact the wire as it was conveyed from drum 2 to drum 3. Sound modulated the current in the electromagnet, 4, to selectively magnetize the wire. The strength of the magnetic flux produced in each part of the wire was proportional to the amplitude of the applied sound. The wire was then rewound and run through a second time. An output electromagnet, 5, similar in construction to the electromagnet (4) generated tiny electric signals as the magnetized wire moved past its coil. These signals were strong enough to activate earphones and to reproduce the sound. To achieve this degree of success, the inventors following Poulsen needed to overcome several problems. It was necessary to design special recording and reproducing heads having very narrow openings. If a signal were of a frequency high enough so that it went through its entire cycle during the time that its corresponding segment of wire was inside the gap, it would not be accurately recorded on the segment; high pitch sounds would thus be lost. The inventors needed also to devise a method for assuring constant wire speed past the heads. Constant drum speeds would not do because the rate of wire feed changes as the take-up reel diameter increases. They were required, in addition, to design a brake system to prevent the feed reel from coasting, causing "spillage," when wire feed was stopped. These provisions were all made but the invention still needed considerable improvement. At this point, inventive interest seemed to lag. In 1943, however, a young inventor named Marvin Camras made several important contributions to the Poulsen concept. Camras took advantage of a highly developed electronic technology available to him (but not to Poulsen) and added tube amplification to both recording and playback functions. This greatly extended the sensitivity, frequency, and amplitude ranges of the apparatus and also allowed certain nonlinearities to be eliminated. He was able to use sophisticated "playback equalization" techniques in the amplifier design to overcome distortion produced by the magnetic heads. A further problem occurred at low levels of signal current. It was difficult to initiate magnetization at these levels; the result was a disproportionate fall off in the reproduction amplitude at low sound levels. Camras added an alternating current "bias" to the recorder head. The frequency of the bias was too high to be recorded but it agitated the magnetic "domains" of the wire, making it possible for small signal currents to create magnetic patterns more in proportion to their amplitudes. The wire recorder was now developed to the point where it could be produced commercially. Many of the recorders were used by the U.S. Navy during World War II and then in the post war years by the general public. This use pointed up further shortcomings in the device and many inventors now turned their attention to these problems. The wire was magnetized in one orientation (Fig. 3-3B) but could easily twist and be presented to the playback head in another (Fig. 3-3C). The results of twisting were frequency and amplitude distortion. These were tolerable in the early uses of the wire recorder for voice transcription, but became undesirable when music began to be recorded. One improvement shown in the patent literature eliminated twisting by flattening the wire into a metallic ribbon (Fig. 3-3D). A further difficulty was found in reproducing the higher frequencies needed for music. In a relatively thick wire or ribbon, too much material is presented for magnetization. Figure 3-3E shows how a later invention overcame this problem. A coating of a thin magnetic iron oxide, 8, placed onto a sturdy plastic strip, 7, permitted the recording and playback of much higher frequencies. Later patents added multiple recording paths (Fig. 3-3F) and moving heads (Fig. 3-3G) to achieve greater and greater recording densities per foot of tape. Along with these developments have been patents covering methods for manufacturing the tape, formulas for magnetic coating compositions, recording circuits for utilizing frequency modulation, pulse coding techniques, and tape handling methods involving cartridges and cassettes. The evolutionary advances contributed by successive inventors in the wire/tape recorder example can be seen as a series of problem-solving steps, each made possible by the previous work. The effect of wire orientation would not have been a noticeable shortcoming if the maintaining of constant wire speed had not been solved first. The non-linearity of magnetization in thick layers of metal would not have been overly objectionable had playback equilization not been first achieved. An inventor desiring to enter the tape recorder field now could extrapolate from present limitations. He could, for example, develop an apparatus with no moving parts whatsoever; a laser or electron beam would scan a card coated with a sensitive material. Modulation of the beam might record the input signal. Rescanning would provide playback. Each card, although small in size, would have sufficient density to store an hour or more of sound or T.V. programming.

We have spoken above of the "sense of knowledge" imparted by the reading of patents. To further illustrate this point, we have selected two illustrations of electrostatic separating equipment. Figure 3-4 is from an engineering handbook; Fig. 3-5 is a sheet of drawings from a patent. Some powdered mixtures can be separated into several fractions according to the relative electrical conductivity of their constituents. A rotating, conducting drum (Fig. 3-4) is fed the powdery mixture from a hopper. A high voltage electrode discharges toward the drum, which is electrically grounded. Nonconducting particles (open circles) become electrically charged, adhere to the drum, and are carried around with it. They are finally brushed off to fall into the end bin. The conducting particles acquire a charge very rapidly but lose it quickly to the grounded drum and are carried by the latter into the first bin. The method can be used to separate metals such as gold, silver, and copper from impurities such as quartz and rock. The patented apparatus of Fig. 3-5 utilizes this

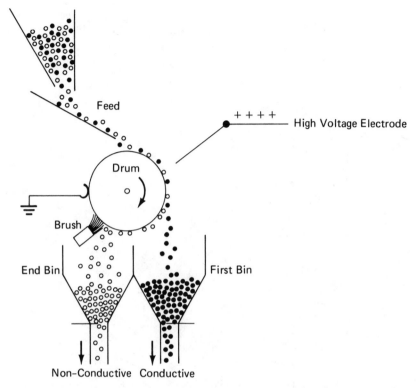

Fig. 3-4. Electrostatic separator principle.

electrostatic principle. The detail of the patent drawings is particularly striking. The electrode, 20, is seen to be in the form of a screen. Powder from the hopper, 12, is carried by the conveyor belt, 22, over the electrode, 14, where the various particulates become charged at differing rates. The rapidly charged conducting material is pulled toward and through the screen by the forces of the latter's opposite polarity and by a vacuum created by the blower unit, 41. The powder-air mixture entering the vacuum system is separated by the centrifugal device, 38. The non-conducting, slowly charged powder, 24, adheres to the belt, 22, and is scraped into the collection bin, 25, by the rotating brush, 26. The handbook illustration and explanation permits the reader to grasp the underlying principle; the patent enables him to start thinking about the design and construction of a workable machine.

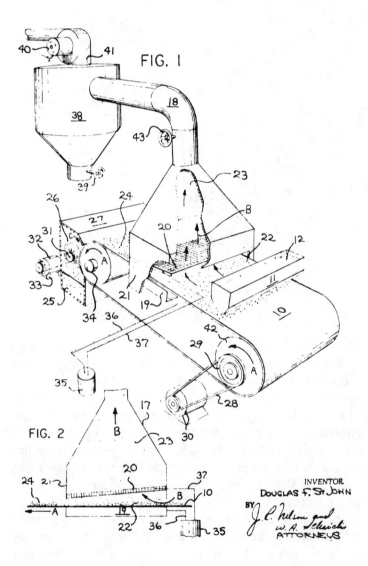

Jan. 13, 1970   D. F. ST. JOHN   3,489,279
PARTICULATE SEPARATOR AND SIZE CLASSIFIER
Filed Dec. 9, 1966

Fig. 3-5. Electrostatic separator patent.

How to Conduct a Patent Search

The term "patent search" refers to a formal inspection of the patent files in order to determine whether a particular invention has been anticipated by anyone else. The task of surveying over 4 million domestic patents (and 20 million foreign patents, to do a really complete job) would obviously be a staggering load for even a large corporation. What must be done, therefore, is to weed out large numbers of non-applicable documents, saving for consideration only those having direct bearing on the subject matter at hand, and selecting for study only a small fraction of the patents dealing with important aspects of the particular invention. To facilitate this process, the Patent Office issues an *Index to Classification,* a loose-leaf volume which lists the major subject headings into which patents have been divided. The major headings, called classes, are presented alphabetically and are assigned numbers (see sample page, Fig. 3-6). It is important to determine all classes under which the subject matter might be listed. A bow and arrow, for example, could be considered as an item under sporting goods, hunting equipment, military gear, stage equipment, toys, line throwing equipment, archery gear, etc. All applicable class numbers should be recorded. The searcher next proceeds to another loose-leaf volume, called the *Manual of Classification,* which lists the classes in numerical order. Under each class number is a group of subclasses more closely defining the specific area covered by the invention. An example will clarify the procedure. Assume that the invention pertains to a radio-controlled signal apparatus to permit a dispatcher to warn a far off train about a dangerous road condition. The *Index to Classification* is found to contain the following pertinent classes: "Telegraph Space Induction," "Train Dispatching," "Signals," and "Railway Switches and Signals." The class number of "Railway Switches and Signals" is found to be 246 (in Fig. 3-6 under "Railways"). In Fig. 3-7 is a page from the *Manual of Classification,* headed by this number and title. Various subclasses appear in numerical order. The subclasses of possible interest here might be 13, 15, 29, 30, and 44. The complete class numbers to be investigated under "Railway Switches and Signals" would be, therefore, 246-13, 246-15, 246-29, etc. It will be recalled that the class and subclass numbers appeared in the title block of the patent studied earlier (Fig. 3-1A).

	Class	Subclass
RADIOACTIVITY		
Contamination detector	250	83+
Demonstrator	250	42+
Cloud chamber	250	
Molding	264	21
Electrically	250	83.3+
Fluorescent	250	71+
Geiger type system	250	83.6
Nuclear reactions involving	176	
Producing in material	176	10+
Neutron type system	250	83+
Photography	250	65+
Scintillation	250	71
RADIOCHEMICAL (SEE RADIANT ENERGY)		
Apparatus chemical change	204	193
Compositions radioactive	252	301.1
Food treatment		
Preservation	99	221
Vitamin activation	99	12+
Medicine radioactive	424	1
Nuclear reactions	176	
Processes chemical change	204	157.1
Radioactive applications	250	106
Sound recording and reproduction	274	5
Tobacco treatment	131	121
RADIOGRAPHY	250	65
RADIOISOTOPE POWERED GENERATOR	310	3
RADIOMETERS		
Light meters	356	213+
Actinometers	95	10
Pyrometers	356	43+
Radio wave meter	250	39
Rotating vane type	356	216
Thermometers	73	355
Ultraviolet	250	83
X-ray	250	83
RADIOSONDE	340	345+
System	340	177+
RADIOTELEGRAPHY		
Earth transmission systems	325	28
Systems	325	26+
RADIOTELEPHONY		
Earth transmission systems	325	28
Systems	325	
RADIUM (SEE RADIOACTIVE)		
RADIUS ROD		
Brake	188	191
Spring braces	267	66+
RAFTER	52	92+
RAFTS AND RAFTING		
Life rafts	9	11
Rafting and booming	9	15+
RAIL (SEE RAILWAY)		
Anchors	238	315+
Bed	5	
Camp	5	117
Elements and details	5	286+
Extension	5	184
Folding	5	177
Rotating	5	302
Benders	72	210
Portable	72	210
Billiard table	273	8+
Bonds for electric railways	238	14.05+
By electric weld	219	53+
Manufacture and installation	29	470+
Chalk and eraser	35	67
Circuit		
Power	191	
Signal	246	
Cleaner	15	54+
Clearer	104	279+
Snow	37	17+
Clip	238	378

	Class	Subclass
Drills portable	408	
Fasteners	238	310+
Fence	256	59+
Fissure detector	73	146
By abrasion	73	8
Electrical	324	37+
Radiant energy	250	83+
Grinders	51	178
Guard		
Railway surface track	238	17+
Track	256	14+
Joints	238	151+
Land vehicle		
Dashboard	296	71
Top	296	123
Manufacture	72	
Curving	72	210
Hard facing	117	
Punching	83	
Razor		
Form	30	83
Guard	30	80
Seats	238	264+
Shapes	238	122+
Shifters	254	43+
Ship belaying pin	114	218
Textile making apparatus	57	136+
RAILWAYS (SEE RAIL)	104	
Chair making	29	16
Draft appliances	213	
Marine	61	67
Pit scale	177	134
Rail or tie shifters	254	43
Rolling stock	105	
Design	d66	
Sounding toy	46	113
Ship	61	67
Signals	246	
Block system	246	20+
Surface track	238	
Electrical connections	238	14.1+
Fence or guard rails	238	17+
Leveling and spacing gauge	410	38
Rail bond making	29	470+
Spike pullers	254	18
Switches and signals	246	
Toy simulation	46	216+
Wheels and axles	295	
Axle making	72	
Wheel making	29	168
RAIN		
Gauges	73	171
Producers	239	14
RAINCOATS (SEE HOIST)	2	87
RAISERS (SEE HOIST)		
Corpse in coffin	27	12
Dough		
Baking powder	99	95+
Heater	126	281+
Jacks	254	
Track	104	7
Wick	431	304+
Wick	431	315+
RAISING SHIPS	114	44+
RAKES		
Agricultural implements	56	
Blanks and processes for making	76	111
Bundling combined	56	341+
Cutter and ground rake	56	193
Cutter with conveying rake	56	158+
Cutter with detachable rake	56	4
Design	d439	1
Dies for making	72	470+
Grain separator	130	

Fig. 3-6 Sample page from the *Index to Classification.*

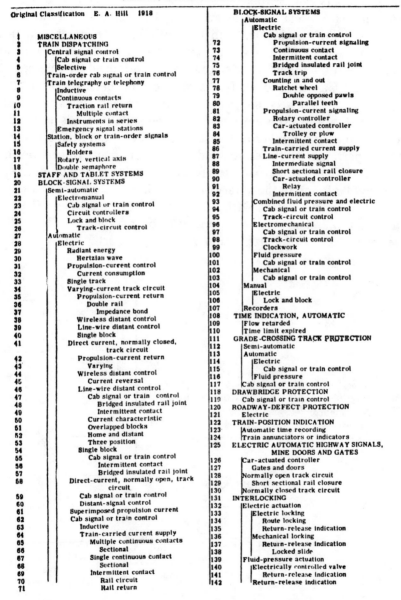

CLASS 246, RAILWAY SWITCHES AND SIGNALS

Original Classification   E. A. Hill   1918

1	MISCELLANEOUS	
2	TRAIN DISPATCHING	
3	Central signal control	
4	Cab signal or train control	
5	Selective	
6	Train-order cab signal or train control	
7	Train telegraphy or telephony	
8	Inductive	
9	Continuous contacts	
10	Traction rail return	
11	Multiple contact	
12	Instruments in series	
13	Emergency signal stations	
14	Station, block or train-order signals	
15	Safety systems	
16	Holders	
17	Rotary, vertical axis	
18	Double semaphore	
19	STAFF AND TABLET SYSTEMS	
20	BLOCK-SIGNAL SYSTEMS	
21	Semi-automatic	
22	Electromanual	
23	Cab signal or train control	
24	Circuit controllers	
25	Lock and block	
26	Track-circuit control	
27	Automatic	
28	Electric	
29	Radiant energy	
30	Hertzian wave	
31	Propulsion-current control	
32	Current consumption	
33	Single track	
34	Varying-current track circuit	
35	Propulsion-current return	
36	Double rail	
37	Impedance bond	
38	Wireless distant control	
39	Line-wire distant control	
40	Single block	
41	Direct current, normally closed, track circuit	
42	Propulsion-current return	
43	Varying	
44	Wireless distant control	
45	Current reversal	
46	Line-wire distant control	
47	Cab signal or train control	
48	Bridged insulated rail joint	
49	Intermittent contact	
50	Current characteristic	
51	Overlapped blocks	
52	Home and distant	
53	Three position	
54	Single block	
55	Cab signal or train control	
56	Intermittent contact	
57	Bridged insulated rail joint	
58	Direct-current, normally open, track circuit	
59	Cab signal or train control	
60	Distant-signal control	
61	Superimposed propulsion current	
62	Cab signal or train control	
63	Inductive	
64	Train-carried current supply	
65	Multiple continuous contacts	
66	Sectional	
67	Single continuous contact	
68	Sectional	
69	Intermittent contact	
70	Rail circuit	
71	Rail return	

	BLOCK-SIGNAL SYSTEMS	
	Automatic	
	Electric	
	Cab signal or train control	
72	Propulsion-current signaling	
73	Continuous contact	
74	Intermittent contact	
75	Bridged insulated rail joint	
76	Track trip	
77	Counting in and out	
78	Ratchet wheel	
79	Double opposed pawls	
80	Parallel teeth	
81	Propulsion-current signaling	
82	Rotary controller	
83	Car-actuated controller	
84	Trolley or plow	
85	Intermittent contact	
86	Train-carried current supply	
87	Line-current supply	
88	Intermediate signal	
89	Short sectional rail closure	
90	Car-actuated controller	
91	Relay	
92	Intermittent contact	
93	Combined fluid pressure and electric	
94	Cab signal or train control	
95	Track-circuit control	
96	Electromechanical	
97	Cab signal or train control	
98	Track-circuit control	
99	Clockwork	
100	Fluid pressure	
101	Cab signal or train control	
102	Mechanical	
103	Cab signal or train control	
104	Manual	
105	Electric	
106	Lock and block	
107	Recorders	
108	TIME INDICATION, AUTOMATIC	
109	Flow retarded	
110	Time limit expired	
111	GRADE-CROSSING TRACK PROTECTION	
112	Semi-automatic	
113	Automatic	
114	Electric	
115	Cab signal or train control	
116	Fluid pressure	
117	Cab signal or train control	
118	DRAWBRIDGE PROTECTION	
119	Cab signal or train control	
120	ROADWAY-DEFECT PROTECTION	
121	Electric	
122	TRAIN-POSITION INDICATION	
123	Automatic time recording	
124	Train annunciators or indicators	
125	ELECTRIC AUTOMATIC HIGHWAY SIGNALS, MINE DOORS AND GATES	
126	Car-actuated controller	
127	Gates and doors	
128	Normally open track circuit	
129	Short sectional rail closure	
130	Normally closed track circuit	
131	INTERLOCKING	
132	Electric actuation	
133	Electric locking	
134	Route locking	
135	Return-release indication	
136	Mechanical locking	
137	Return-release indication	
138	Locked slide	
139	Fluid-pressure actuation	
140	Electrically controlled valve	
141	Return-release indication	
142	Return-release indication	

Fig. 3-7 Page from the *Manual of Classification*.

Once the class and subclass numbers are determined, the actual patent numbers can be found from indices appearing as yearly summaries in the Official Gazette and on microfilm in some libraries. These indices relate class and subclass with patent numbers. At this point, the searcher will have acquired a relatively large number of specific patent references. He now refers to individual volumes of the Official Gazette and reads abstracts of the patents. From the abstracts and their accompanying drawings he eliminates the references which are of no importance. The remaining patent references have now passed several critical tests. They are of the title, class, and subclass pertinent to the search and in abstract appear to bear on the matter at hand. These patents are now obtained in their entirety. They may be found in the library or may be individually purchased for 50¢ each from the Patent Office. From the references given in each patent, the searcher may find previously overlooked patents or discover new classes or subclasses under which prior art is listed.

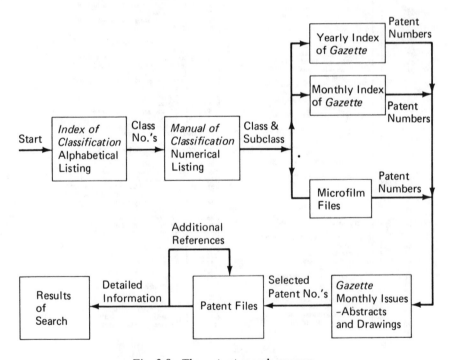

Fig. 3-8. The patent search process.

The search process is diagrammed in Fig. 3-8. The procedure may be somewhat arduous for the beginner but becomes relatively routine with practice.   Every inventor should make at least one search in a year in order to maintain his familiarity with the patent files.   When a professional search is ordered through a patent attorney or agent, the results are often provided in the form of a recommendation to the inventor, who should ask for the key patents, at least, and study these.

The question is often asked by inventors whether a search provides the last word on the uniqueness of an idea. Unfortunately, it does not. In the huge collection of patents, both in the United States and in foreign countries, there is often some overlooked item which has direct bearing on the idea under consideration.   The examiner conducts a much more thorough search when processing an application and will frequently turn up prior art which has been missed by the inventor.

Libraries which Contain Complete Sets of Patents

City	Library
Albany, New York	University of the State of New York
Atlanta, Georgia	Georgia Tech Library
Boston, Massachusetts	Public Library
Buffalo, New York	Buffalo and Erie County Library
Chicago, Illinois	Public Library
Cincinnati, Ohio	Public Library
Cleveland, Ohio	Public Library
Columbus, Ohio	Ohio State University Library
Detroit, Michigan	Public Library
Kansas City, Missouri	Linda Hall Library
Los Angeles, California	Public Library
Madison, Wisconsin	State Historical Society of Wisconsin
Milwaukee Wisconsin	Public Library
Newark, New Jersey	Public Library
New York, New York	Public Library
Philadelphia, Pennsylvania	Franklin Institute
Pittsburgh, Pennsylvania	Carnegie Library
Providence, Rhode Island	Public Library
St. Louis, Missouri	Public Library
Stillwater, Oklahoma	Oklahoma Agricultural & Mechanical College

| Sunnyvale, California | Public Library |
| Toledo, Ohio | Public Library |

Other Sources of Information

Even if an invention has not been patented by someone else prior to the time that the application is sent in, the inventor may still be barred from obtaining a patent. If the idea has been described in a publication which appeared earlier than one year prior to the patent application, the concept is considered to be in the public domain. In this case, no one is entitled to the exclusive right to "manufacture, sell, or use" which a patent confers. To properly investigate all the literature would require the searcher to read thousands of journals, magazines, and newspapers in many languages — a somewhat hopeless task. Fortunately, there are a large number of digests which condense most of the world's technical literature and make it feasible for an individual to cover large amounts of information in a relatively short time. A listing of some of these follows.

Abstract Journals
(Condensations of articles which have appeared in world technical literature)

* 1. *Applied Science and Technology.* An alphabetically-arranged summary of many sciences. Issued monthly. H.W. Wilson & Co.
* 2. *Engineering Index.* A monthly summary covering the engineering fields. Engineering Index, Inc.
** 3. *Chemical Abstracts.* Comprehensive listing of progress in chemistry, chemical engineering, pharmacy, agricultural chemistry, etc. American Chemical Society.
** 4. *Physics Abstracts.* A summary of current work in such areas as optics, heat transfer, atomic energy, and electrical theory. Institution of Electrical Engineers.
* 5. *Geophysical Abstracts.* Current work in earth sciences. Geophysical Survey.

Encyclopedias

* 1. *Kirk-Othmer Encyclopedia of Chemistry.* Alphabetically arranged. Interscience.
** 2. *Encyclopedic Dictionary of Physics.* Macmillan.

*Moderate level of technical comprehension required.
**Highly technical.

Many other summaries and encyclopedias of world technical literature (at all levels of complexity) are available in libraries.

As a final word in regard to the searching of the patent literature: it is often very discouraging for an inventor to find prior art. He begins to feel that he has wasted his time and effort. A highly experienced and competent patent attorney once commented on this to the author. "The only time that some hint of identical prior art is *not* found in a good search is when the inventor is either an outstanding genius or if he is working in a completely valueless field." This man felt that the finding of prior art should be regarded as an encouraging sign that the inventor is on the right track and should by all means continue his work.

4

The Inventive Process

In Chapter 1 were listed the most frequent types of invention:

1. Single or multiple combination
2. Labor saving devices
3. Direct solutions to problems
4. Adaptation of an old principle to bring about a new result
5. Application of a new principle to an old use
6. Application of a new principle to a new use
7. Application of an accidental discovery

Although a listing such as this permits one to classify many patents and to better comprehend the objectives sought in each case, it would only be of preliminary help in establishing an inventive procedure. What we desire here is a basic "how to" approach to the business of inventing. How does one get original ideas, how does one see through to the heart of a problem, how does one devise a mechanism or system to achieve a particular result? The mental processes which preceded or brought about great inventive achievements have been studied for years. By questioning the individuals concerned, investigators have drawn a number of generalizations as to what is required to produce a great invention, a new scientific theory, or a solution to an important problem. Using these findings, researchers have devised systems for aiding the creative process and for increasing the yield of innovative ideas. In this chapter, we shall discuss a number of these systems.

Preliminary Considerations

Innovating is, in general, a problem-solving activity. The individual finds himself separated from some goal by one or more barriers. His problem is to overcome each barrier and reach the goal. If we symbolically represent the barrier as a physical wall preventing us from continuing along a certain path, we can stipulate that only four methods are available for overcoming the wall (Fig. 4-1). We can break directly through the barrier via path A, we can go over it by path B, under it using path C, or around it via path D. The direct path A is usually the most difficult. Any current state of the art has usually been arrived at by a previous series of direct assaults. It becomes increasingly harder to directly improve a thing which has been improved many times before. Paths B, C, and D, on the other hand, require more ingenuity. A specific example will illustrate these four approaches. Assume that we wish to design a gasoline driven turbine for use in automobiles. Initial studies indicate that a turbine engine must be run at a constant high speed if it is to operate efficiently. Methods for gearing down and then varying this high speed are available. The high temperatures produced by combustion and the high

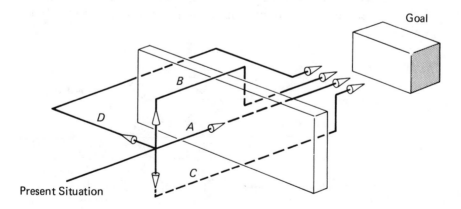

Fig. 4-1. The four general approaches to problem-solving.

centrifugal forces causes destructive creep effects on the metal of the rotor. In addition, the impact of hot combustion gases on the rotor blades causes severe erosion. In tests in which a turbine-engine car was driven for 10,000 miles, the turbine blades became so unbalanced that the main bearing was damaged. How can this problem be overcome? The direct approach would attack the problem metallurgically. Attempts would be made to invent an alloy having extremely low creep at high temperatures and good resistance to erosion while still retaining such desirable features as ease of fabrication and high tensile strength. A path *B* approach would be to provide cooling ducts inside the turbine and arranging for high velocity flows of coolant within the moving blades. This would require the development of rotary seals able to operate at high speeds, temperatures, and pressures. The turbine's mechanical output could be tapped to drive the coolant pumps. A path *C* solution would involve the use of two or more turbines driving the same shaft. Hot gases would be distributed on a time-sharing basis to the various turbines (similar in principle to the manner in which the cylinders of a piston type engine are fed). During the one or more revolutions that a turbine was not being fed hot gas, it would pump air and thus be cooled prior to its next "on" period. A path *D* approach would modify the turbine concept; a reaction wheel similar in principle to a rotary lawn sprinkler would be substituted. Hot combustion gases would be admitted at the center of the wheel through a rotary seal and exit through the ends of the curved arms. Impact erosion on moving parts would be reduced. The inside passages of the arms would be lined with high temperature ceramic, and cooling air would be added to the housing (Fig. 4-2).

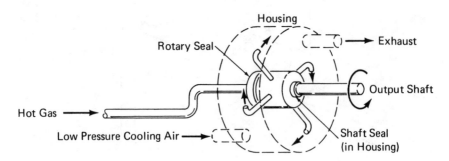

Fig. 4-2 Rotary engine.

The term "creativity enhancement" is used to describe those procedures whose aim it is to make problem solving in all areas — science, engineering, business, politics, arts, etc. — easier and more direct. In the next few sections, we shall describe some of these schemes and indicate their strengths and weaknesses.

All creativity enhancement methods are based on overcoming certain behaviour patterns. Under normal conditions, the mind and body find it efficient to form habit responses to various stimuli. The many routine operations associated with our daily lives can thus be carried out with minimum effort. When some unusual problem occurs, there usually arises a desire to dispose of the interruption to our routine as quickly and painlessly as possible. This response is not consistent with a high degree of creativity. Some escape mechanisms by which the mind rebels against unaccustomed activities are very familiar: avoidance, procrastination, diversion, distraction, etc. We tend to avoid acting on the problem altogether, putting it off, becoming interested in less relevant aspects, or seeking other pursuits. Certain creative thinking techniques attempt to convert the work of innovating into game activity and thus make use of the trend towards diversion. Another response to departure from habit is to fall back on authority or established practice. By setting up artificial conditions, some creativity methods arbitrarily remove these props and permit a more flexible exercise of the imagination.

Problem Definition and Problem Assignment

An initial stumbling block in all creative efforts is the proper definition of the problem itself. It is not hard to see how an innovative situation can become somewhat vague. An inventive individual reads about a new kind of fluid which is attracted by a magnet. This property is exciting; he feels that there must be a number of uses which are unique and patentable. After some thought, he devises a mechanism in which the fluid is confined in an elastic bag along with an internally positioned electromagnet. When the latter is energized, the bag bulges at one end and thus imparts movement to a nearby rod. The inventor utilizes this motion to close electrical contacts. He constructs a rough working model to verify operability. He has, in effect, invented a different type of relay. During all of this, a number of objectives needed definition but were put aside in the exhilaration of new discovery. Some of these were:

1. What end result have I been trying for? Is it just to find an application for the new liquid?
2. What is the present industrial situation with regard to relays? Are new mechanical types being sought for any particular purpose?
3. What refinements are required? (To meet present day operating speeds, for instance.)
4. What tests will I apply to study the characteristics of the device?
5. Suppose everything works out well (from the device stand-point), what will I do with it next?
6. What other projects am I now engaged in and will the work on this development take priority?

The inventor has taken a number of steps before having considered even in general terms what he wants to accomplish. As a result, he may find himself with an unmarketable product protected by a valid patent. The above statements are in no way a condemnation of free association or unimpeded use of the imagination. What is required, however, is some initial set of rules for each project by which to guide one's thinking, eliminate the less viable ideas, make the optimum use of one's time and properly exploit the finished product.

One significant difference between the intellectual environment of the industrial scientist and that of the independent inventor originates from the fact that the former "invents to order." He is presented with a problem by his employer along with some set of working rules. These define the area of endeavor and narrow down the range of possible approaches. The industrial man does not need to worry about the eventual use of any discovery he might make. The problem assigned may be in one of several forms:

1. *A specific product improvement.* A razor manufacturer, for example, instructs his research department to devise a new shaving implement which will hold six blades internally and be disposable after the last blade is used. Blade angle shall be adjustable by the user.

2. *Solving a perceived problem.* An instrument manufacturer who makes smog measuring equipment wants to diversify. He directs his researchers to devise an explosives detector for checking passenger luggage at airports.

3. *Developing an overall system.* A governmental agency awards contracts to several research organizations to originate and develop an early warning network for tracking and destroying multiple-headed rocket vehicles approaching this country in an undeclared war situation.

4. *Devising a short term problem remedy.* A manufacturer of high quality plate glass has for years dumped partially fused sand onto a large field adjacent to his plant. His chemists are now directed to find ways of utilizing this waste product in some profitable manner. If a process is found and the material disposed of, it may be necessary to purchase similar waste from other manufacturers or allow the development to lie idle until more material accumulates.

5. *Good will problem solving.* A brewing company has had a particularly good record of profitability. To improve its public image, this company directs several of its research personnel to devise new methods for producing edible protein from seaweed. The results are made available without charge to under-developed countries.

It will often be profitable for the independent inventor to emulate the industrial scientist and introduce more objectivity into his own efforts. He should outline the problem, define the methods by which he wants to attack it, and list what steps to take after each kind of result. He should even plan on how he will close out the project — abandon it in two years if the effort proves unsuccessful, patent the result, try to sell the rights, donate the results to the public, start a manufacturing operation, etc.

The Enhancement of Creativity

Listing Methods. The simplest technique for systematic idea generation is to make a list of the attributes of the problem. This forces a breakdown of a complex subject into a number of more easily handled subdivisions. Each of the latter is then examined and "free thought"

TABLE 4-1. ATTRIBUTES OF PRESENT DAY WINDMILLS.

Item No.	Name	Comments
1	Tower	Four-sided steel skeleton. Must allow wind to pass through easily. May be necessary to use guy wires with large structures. Optimum height?
2	Blading	Multi-blading used in older styles. Only three blades in modern version. Why? Provision for automatic "feathering" in high winds. Is there any way to avoid feathering and utilize all the wind power during a gale? Is entire force of the wind utilized? Suppose we used two or more sets of blades in tandem, or blades operated in some kind of wind-directing tunnel?
3	Orientation device	Tail fin keeps blade facing wind. This complicates power take-off systems. What is current thinking on vertical shaft devices which operate equally well with winds from all directions? Is wind always horizontal? Should blading also tilt to take advantage of ascending or descending winds?
4	Power production system	Older mills used bevel gears, which transmitted shaft work down the middle of the tower. Newer, small units employ automobile alternators. Can wind energy be converted to any other form for storage? Decompose water to hydrogen and oxygen and store below ground? Possible use would be on North Africa coast to make potable water from the sea. What is optimum speed for present day generators? Should blade output be geared?
5	Base	Present devices use cast concrete footers. Are there applications where windmills might be mounted on mobile platforms to be relocated with the seasons?
6	Miscellaneous	Present diameters seem limited to fifteen feet. One large mill built years ago in Vermont was destroyed in storm. Present strength of materials technology probably improved enough to prevent this from happening now. What about "multiplicity" factor? Are three 15 ft. windmills equivalent to one 45 foot unit? How to determine optimum diameter from efficiency, economics, and durability standpoints?

notes made concerning that part or item. As an example, suppose we wished to invent a new kind of windmill which would be suitable for large scale production of power. We have examined an existing model and have listed components in the left hand side of Table 4-1. Comments and questions about each item have been entered in the right hand column.

As the problem of windmill design is broken up into smaller units and different components of each unit are considered, questions arise as to why certain design decisions were made in the past. Ideas for new investigations and improvements suggest themselves. The listing is not necessarily a "one sitting" project; comments can be added or deleted over any desired period of time. "Flash" ideas should be entered as soon as possible before they are forgotten. Contributions by others can be in the form of direct addition or by adding on separately made listings.

A variation of the listing method involves treating the problem and possible solutions as though they were special items or services being ordered from a supplier. This method has been called the input-output, or specification, technique. All phases of the problem and all aspects of the desired solution are described in detail. As an illustration, assume that our problem is smog elimination from cities. We prepare a listing as shown in Table 4-2.

TABLE 4-2 SMOG ELIMINATION.

Characteristics: Semi-visible vapor containing carbon dioxide, carbon monoxide, various nitrogen oxides (one of which is nitrogen dioxide, a reddish brown gas), sulfur dioxide, mercaptans, fine particles, and miscellaneous hydrocarbons. Composition varies with sources, climate. Many of the components are poisonous.

Remarks: How are the gases measured? What are toxic limits of the poisonous components?

Origins: Attributed mainly to auto exhausts. Contributions by refineries, steel plants, coal burning installations.

Remarks: What chemical reactions are involved in the origin? Can any of these be reversed as a means of elimination?

Specification of the Remedy: Must be applicable to autos. Must not generate undesirable by-products or damage equipment. Must be relatively inexpensive.

Remarks: Can gases be pumped directly out of the air? e.g., suction fans alongside freeways? What gasoline additives could be considered?

Matrix Methods

If a device or system is made up of only a few critical parts and each of these can be chosen from a relatively small number of possibilities, each and every combination can be determined. An all-inclusive test program will then decide the optimum combination.

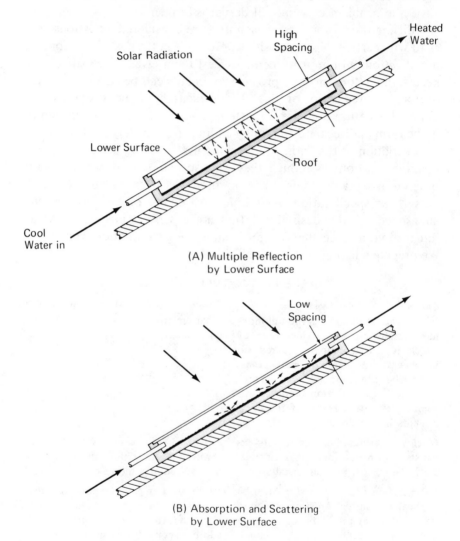

Fig. 4-3 Solar-powered heater.

A matrix or mathematical table is first set up to indicate the variables. One example would be the solar heating device shown in Fig. 4-3. It is comprised of a glass covered rectangular chamber to be placed on a roof. The sun's rays enter the glass and impinge on the bottom of the chamber. Here they are either reflected, absorbed, or scattered. Water is pumped through the chamber, becomes heated and is then circulated through radiators within the house. Initial studies and experiments show that the most important variables are the color of the lower surface, the texture of that surface, and the depth of the chamber. If the lower surface is shiny, multiple reflections are obtained between the two inner surfaces of the chamber. This may aid in the maximum utilization of the light. If the lower surface is rough and light in color, radiation is scattered in all directions. Some may escape back through the glass cover but a high percentage is re-reflected and dissipated as heat in the walls of the chamber. If the lower surface is rough and of a dark color, most of the radiation will be absorbed there and heat the bottom of the chamber. How much of this heat is transferred to the water will depend on conduction effects and losses by re-radiation. The closer the spacing, the greater the number of reflections. With this information we can now prepare a matrix containing the variables and design an experimental program. The matrix is shown in Table 4-3.

TABLE 4-3. SOLAR HEATER VARIABLES.

Color of lower surface	White (A_1)	Silver (A_2)	Black (A_3)
Texture	Shiny (B_1)	Rough (B_2)	—
Spacing	High (C_1)	Low (C_2)	—

For convenience, each variable has been given a symbol. It can be seen that there are 3 x 2 x 2 or 12 possible combinations. These are:

$$A_1 B_1 C_1 \qquad (1)$$
$$A_1 B_2 C_1 \qquad (2)$$
$$A_1 B_1 C_2 \qquad (3)$$
$$A_1 B_2 C_2 \qquad (4)$$
$$A_2 B_1 C_1 \qquad (5)$$
$$A_2 B_2 C_1 \qquad (6)$$

$$A_2 B_1 C_2 \qquad (7)$$
$$A_2 B_2 C_2 \qquad (8)$$
$$A_3 B_1 C_1 \qquad (9)$$
$$A_3 B_2 C_1 \qquad (10)$$
$$A_3 B_1 C_2 \qquad (11)$$
$$A_3 B_2 C_2 \qquad (12)$$

Each experiment is now run and the results compared to find the best combination. All possibilities can be tested. Unfortunately, as the number of variables increases, the number of combinations grows rapidly. If there are four variables, for example, each having three possible conditions, the possible combinations total 3 x 3 x 3 x 3 or 81. It is sometimes feasible to eliminate certain combinations, on the basis of experience or calculation. It has been possible in large, well-funded research investigations to set up computer programs which "run" experiments in simulated form and thereby try out all the combinations. The independent inventor will often find matrix methods valuable for idea stimulation, even when the number of experiments required for total evaluation is too high for practical purposes.

Changing Viewpoints Method

It sometimes happens that an investigation becomes stalled at some point and no further progress seems possible. Conventional attack appears to suggest no way of proceeding further. A method which is often profitable in these cases is to change one's viewpoint. If, for example, a new type of tire tread for use in snow is being sought, the inventor can imagine himself to be the road surface and try to visualize what happens when he encounters a tire. He may find so many unanswered questions that he builds a glass roadway section and takes high speed pictures from underneath when vehicles are driven over it. It turns out that the snow melts at the point of contact with the tire so that the vehicle is actually traveling on a film of water. The improved tire must provide means for eliminating this film. In another case of viewpoint changing, a professor of creative design has his students imagine a planet on which the inhabitants have three eyes, three fingers on each hand, breathe methane, and live in a very high gravitational field. The students must then design everyday articles for these people — houses, cars, furniture, utensils, and clothing. Each student becomes, in his mind, one of the dwellers in this strange planet and tries to visualize what would be comfortable, safe, usable, or convenient. The exercise generates idea flexibility and fluency.

A closely allied technique to viewpoint changing is that of reversal. The tail is made to "wag the dog" and the results are analyzed. If the problem were to find a way to relieve highway congestion at rush hours, the inventor might reverse the usual travel process. The highway is visualized as a moving concrete ribbon which brings the work to and away from stationary employees. Further development of this theme involves a conveyor belt system with branches to various neighborhoods so that a worker would only need to travel a few blocks to a "subplant" in his vicinity. At a later stage in the study, a survey might reveal that most of the work done in the densely concentrated downtown area involves the handling of paper work. The adaption of electronic networks and computer might permit each employee to do his work from his own home.

Sometimes reversal will of itself produce a significant improvement with little or no further analysis. Up to the twentieth century, building construction made dual use of walls: to protect against the elements and to hold the building up. This latter function compelled architects to use walls which were very thick at the base in order to support the accumulated weight of the entire structure (Fig. 4-4A). Reversal of the concept, letting the building hold up the walls, led to radically new types of construction. The floors were now supported on cantilevered beams extending from a central column (Fig. 4-4B). This generated the concept of "curtain" walls. In many cases, the walls are almost entirely of glass because great compressional strength is no longer needed.

Another variation of the viewpoint changing technique is that of shifting emphasis. In concentrating on the very center of a problem, we often find outselves blocked and unable to continue. Shifting emphasis from one part of the problem to another may open new paths. A medical researcher is attempting to find a way to measure the force exerted by the heart under various stimuli and when affected by disease. Direct incision and emplacement of a transducer

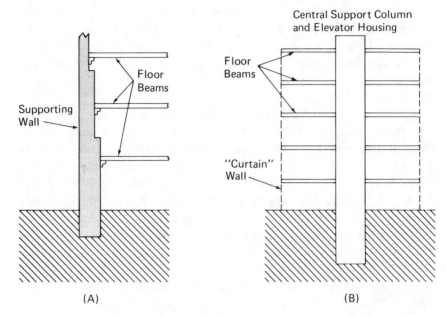

Fig. 4-4. Reversal of function.

is practical with laboratory animals but obviously of limited use with humans. He shifts his attention from monitoring the heart to monitoring the rest of the body. As an experiment, he places a stopwatch on a slightly convex, smooth surface. He notices that the case of the watch moves slightly with every tick — a reaction to the torque produced by the balance wheel. His next step is to build a spring suspended table to which the patient can be strapped. He is delighted to find that the whole body vibrates slightly with each heart beat. It is now a simple matter to measure this vibration and to calculate the force exerted by the heart.

Question Asking Methods

Some inventors have developed a series of questions with which they break down a problem and speed the production of new ideas. A similar technique has been taught for many years to student reporters. The six questions which a reporter must ask are: how, where, when, why, what, and who? When these are answered properly, the reporter knows he has obtained the true essentials of a story and will not be embarrassed later by the absence of some important detail. Problem solving questions have been devised by Osborne and others (see references at the end of this chapter). One set of these is as follows:

1. Can this device be used for other purposes?
2. What older idea can be used here?
3. Will changing shape, color, texture, odor, etc. do anything to solve the problem?
4. How about increasing the size, strength or quantity? What would be the effects of miniaturization?
5. Can a less expensive (or more expensive) component be substituted?
6. Can components be arranged in some new way? Can a new ingredient or catalyst be profitably employed?
7. Can the properties of several materials be combined to give a solution to the problem?

Updating and Adaptation

Thomas Edison recommended that inventors be on the lookout for novel and interesting ideas that others have used successfully. This was not to condone the piracy of mental property but to encourage the application of tried concepts in an original way to the solution of problems. A study of the patent literature often turns up good ideas which are in the public domain. The originator may not have been able to exploit his property during its legal lifetime because suitable materials had not yet been developed or simply because the idea's time had not come. Often an idea long in the public domain can be "dusted off" and presented to a new generation which has no knowledge of its previous popularity. Hoops made of rattan were used by children at the turn of the century to spin around their bodies. The idea reappeared much later in the form of the "hula hoop" but the item was now molded from plastic. Many millions of the "novel" hoops were sold.

An old dream of electrical engineers has been the transmission of power by radiative methods. The advantages would be obvious. A central station would transmit power to the antennas of automobiles. Electric motors would provide quiet and smooth traction to each wheel. Periodic stops for fuel, as well as smog formation, would become memories of the past. Previous inventors (Nicola Tesla, among others) were unable to overcome the great losses encountered when high power is converted to radio frequency energy and radiated into space. Present work with the laser is encouraging in that it is now possible to beam power over considerable line-of-sight distances with much lower dissipation. It would appear that radiative power transmission is an old idea which is now becoming possible because of new technical developments.

Biological Modeling

An extremely powerful method for enhancing creativity is to start by asking one basic question: how is this problem handled in nature? We need only to examine the structure of a blade of grass, an insect, or the skeleton of an animal to find many ideas for ingenious mechanisms — all free for the asking. There are hundreds of cases where a solution to a problem lay unrecognized in nature for many years. One example is a phenomenon observed for centuries — a diseased body buried for a few weeks and then exhumed loses all trace of the disease which killed the victim. The microorganisms seemed to prefer living tissue. It was only after the antibiotic effects of penicillin were discovered that this puzzle was solved. Yet penicillin, produced by mold in the soil, was a well known substance.

It has been long known that uncovered honey will not attract most insects or airborne organisms. Bowls of honey found in ancient tombs have crystallized but can be reconstituted with water and are found to be unspoiled. It remains for some investigator to isolate the resistive factor in honey and to apply it to human health.

Nature has constructed many devices by which its creatures move, navigate, maintain equilibrium, and exert local control over environmental conditions. The high speed of the bottle-nosed dolphin through water is related to the way it ripples its skin surface and somehow decreases friction resistance. Bats can fly through an array of closely-spaced vertical wires even when blindfolded because they emit a series of ultrasonic beeps, the echoes of which are processed by an on-board computing network to give continous range information. A recent application of this principle will be discussed in Chapter 5. The navigation system of certain flying insects has fascinated biologists for a long time. It was discovered that the eyes of some bees are sensitive to the degree of polarization of light. It was also found that polarization occurs in light coming from a northerly direction to a greater degree than that from other directions. It is this difference that is utilized by bees in obtaining direction for navigation. The principle has been applied to a non-magnetic compass for use in the polar regions where ordinary compasses are inoperative.

The flow of blood in the body is metered and controlled by an elegant measuring system. An electrolyte is injected into the blood at some point and its time of arrival at a fixed position from the point of injection is determined by a conductivity detector. The blood is then purified of the electrolyte and is ready for another injection. A computing system determines velocity from the traverse time of the electrolyte-containing blood and then regulates the heart beat. This somewhat simplified explanation is intended only to indicate the elegance of this natural flow measuring system. Many industrial processes use recirculating liquid streams − e.g., boilers, lubricating devices, air conditioning systems − but no application has yet been made of this flow measuring scheme.

Nature is an excellent instructor in the field of adhesives. The bonding between the various building blocks in natrually occurring substances represents a wide range of types and strengths. The principle of interlocking polymer chains to produce solid plastics from liquids has long been used by nature in congealing blood and in producing hard gums from the liquids exuded by trees. Wood consists of long chains of cellulose locked together in rigid form. An adhesives secret still guarded is the material used by certain marine organisms such as the barnacle. This adhesive will harden under water and adhere to almost any surface. The bond is flexible and appears to be thixotropic − it hardens and becomes firmer whenever some force attempts to dislodge the barnacle; when the force eases, the bond softens. Studies were made to isolate this binding substance, the intention being to discover a base for a superior dental cement. These efforts were not successful and have for the time being been abandoned.

Nature abounds, however, in many more easily utilized solutions to inventors' problems. The distribution of seeds from plants is an example. The fall of seeds from trees is slowed by wonderfully contrived aerodynamic devices. A basswood seed is mounted on two winglike structures which impart a spinning motion. The spin slows the fall, thus allowing wind currents time for wide scattering. Perhaps the most interesting falling motion is that of the ailanthus seed, a model of which can be constructed by twisting a thin strip of paper at both ends. If the twists are the same, the strip will spin rapidly about its own length axis and fall along a 45° line. If one twist is greater than the other, two motions will be observed. Not only will the strip spin on its axis as before but it will also follow a spiral path downward. Professor Victor Papanek, Dean of the Design School at the California Institute of the Arts, has proposed using plastic "seeds" to help put out forest fires. The seeds, containing fire-extinguishing powder, would be released by the thousands from airplaines at a point just below the fire-induced updraft. The spiraling paths of the synthetic seeds would direct them to the hottest part of the fire. As they approached, the plastic would melt and release the powder. Other uses suggested for these seed structures are reforestation of desert areas and fish restocking.

Living things have many built-in mechanisms for guarding their well being. Aside from the familiar "fight or flight" options in animals (one of which is rapidly chosen when danger threatens) there are many other monitoring and protection systems that act to overcome destructive forces. Well-known examples of these mechanisms include:

1. The changes in the production rate and distribution of white cells in the blood when infection occurs. White cells increase in number, surround and destroy invading microorganisms.

2. Heat-regulating systems that alter perspiration rate and blood pressure to adapt body temperature to changing ambient conditions and movement.

3. Healing liquids that bring about on-line recuperative effects when an accident or breakdown occurs.

A few man-made devices have simulated these natural mechanisms. Urban electrical systems, for example, are always provided with some standby power arrangement. This may be in the form of a rapid-starting spare generating plant or a switching apparatus connected to the power source of a nearby community. The line voltage is continuously monitored. Should there be a drop, standby power is automatically made available to the load.

Some interesting attempts were made a few years ago to develop self-healing electrical circuits. When a copper wire melted or eroded, a chemical contained in a special chamber within the wire was released and began to react with the metal. The reaction produced crystal growth in the form of long "whiskers" which would bridge the gap and restore operation. The process, unfortunately, was not rapid enough to be practical. A successful method for electrical self-healing would greatly increase the reliability of hard-to-reach systems such as underwater telephone cables and communications satellites. Self-healing processes would also be useful in the hydraulic circuits of aircraft and in the filaments of incandescent bulbs mounted in remote areas.

Analysis and Synthesis

Two somewhat sophisticated inventing techniques are those applying analysis and synthesis. These both depend on a fairly complete knowledge of a particular field and experience in manipulating various components and experimental parameters. In the analysis method, a given device, system, artifact, or natural structure is studied and its underlying principle discovered. The general law is then applied to other problems. The synthesis method consists of learning only the end uses and advantages of a particular device. The device is now "reinvented;" parts and systems known to the inventor are put together (on paper) so that the finished product will do exactly the same job as the original device. Now the makeup of the latter is studied. The inventor compares his synthesis with the original. It will often happen that the synthesis represents an imporvement. If he has not improved the device, the inventor has at least learned its workings in a very short time and is in a good position to continue his development work. Examples of analysis and synthesis will illustrate these methods.

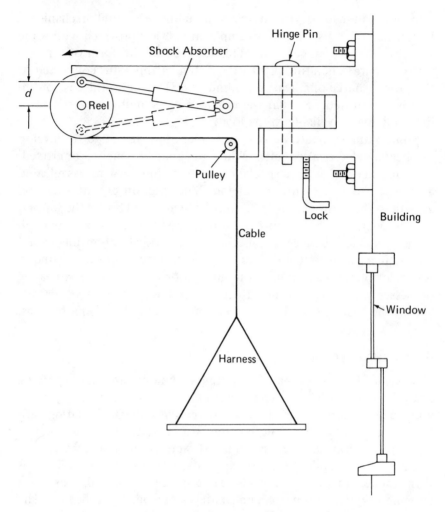

Fig. 4-5. Fire escape apparatus.

The inventor reads about an escape device to be used for persons trapped in a burning building. It is comprised of a long wire cable stored on a reel and a shock absorber of the type used for automobiles, assembled as shown in Fig. 4-5. As the cable is payed out, the shock absorber alternately extends and retracts. Its frictional damping resistance serves to slow down the travel of the harness so that the user is lowered at a safe speed to the ground. The shock absorber can then be disconnected at its pivot and the reel rewound. A more elegant version would provide a clutch and spring return mechanism to rewind the cable. Analysis of the invention would start with a study of how a shock absorber works. It is essentially a spring rigidly coupled to a piston-cylinder arrangement (Fig. 4-6A). If the top

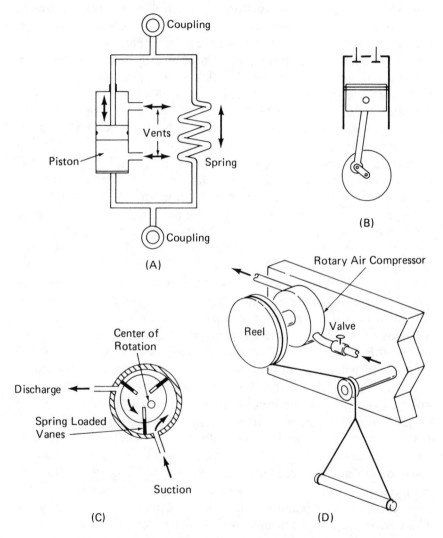

Fig. 4-6. General principle of the fire escape apparatus and subsequent improvement.

coupling is forced downward (the bottom coupling being fixed), the spring will compress but at a rate determined by the escape of air from the space below the piston and the rate at which air is added above the piston. When the top force is removed, the spring will restore the initial condition but also at a rate determined by the movement of air in the spaces above and below the piston. By controlling the orifice sizes of the two vents, the response speed of the shock absorber can be varied over wide limits. The analyst now proceeds further. What other common use is made of this principle? In the power system of the automobile itself, of course! The motorist takes advantage of the damping and braking of the engine every time he lifts his foot from the accelerator. Inertial forces (the flywheel's spin and the forward motion of the car) tend to maintain speed; the restricted movement of gases into and out of the cylinder dissipate energy and cause the vehicle to decelerate. The analyst has now arrived at the general principle of the escape apparatus. Proceeding from this point, he notes that the moment arm (the distance d in Fig. 4-5) acting on the shock absorber changes as the reel turns. As a result, the braking action is least at the top and bottom "dead center" and the operation of the escape apparatus will probably not be smooth. This would probably be uncomfortable for someone being lowered from a multi-story building. Referring back now to the general principle of trapped air exerting strong resistive force, he thinks about piston type air compressors (Fig. 4-6B) and how their pulsations are smoothed out by the use of storage reservoirs. This line of thought leads to a consideration of rotary air compressors. In one form of the latter (Fig. 4-6C), a solid cylinder is mounted off-axis within a larger, hollow one. The smaller cylinder is slotted to house spring loaded vanes which press against the inner walls of the hollow cylinder. As the inner cylinder is rotated, it forms chambers of gradually decreasing volume. This causes air to be brought in from a suction port and discharged through an outlet port. The resistance to rotation is relatively uniform throughout each revolution because of the multiple blading. The inventor now mounts a rotary air compressor on the axis of the reel carrying the escape harness cable (Fig. 4-6D). A valve is added to the inlet port to control rotation speed. The device has now been improved in the following ways: the non-uniform motion is overcome, the apparatus is more compact, and rewinding requires only the opening of a valve (instead of disconnecting the shock absorber at its pivot as in the original model). The compactness and ease of adjustment suggests other applications. The principle might be applied to fishing reels as a means of altering drag. This latter use has a much greater sales potential than the original escape device.

In summary, the process of analysis consists of the following steps:

1. Thoroughly inspect all parts of the device, phenomenon, or problem.
2. Bring to bear experience, literature, calculations, etc. to derive the general, underlying principle.
3. Extend, reapply, modify, or improve to lead away from the general principle to the specific application.

In the synthesis technique, by contrast, detailed knowledge of the present state of the art is carefully avoided. The "little black box" concept is a convenient means for applying the synthesis method. The device is conceived as being a two part assembly: a visible part operating in an obvious way and an invisible mechanism contained in a black box. The synthesis starts by speculations as to what might be housed in the black box in order that the overall result can be obtained. After a number of possibilities have been explored, the inventor now permits himself to learn how the actual mechanism is constructed. If he has exactly matched the latter with one of his suppositions, he has brought himself up to date on the state of the art in a very short time. If his suppositions are more primitive, he has only to learn an incremental amount to achieve current status. If one or more of his suppositions is advanced over current status, he has invented an improvement. A simple example is shown in Fig. 4-7. The item in question is a quick opening valve to be used for putting out aircraft fires. An electrical signal from a fire detector causes rapid opening of a butterfly valve, which then floods various compartments with a foaming material to choke off the flame. The inventor sketches in the obvious parts of the mechanism — the valve, rack-and-pinion operator, and the drive rod — and a black box which contains the unknown. His first speculation is that the end of the drive rod terminates in an iron core and that the box contains a solenoid (Fig. 4-7B). The application of current to the input leads

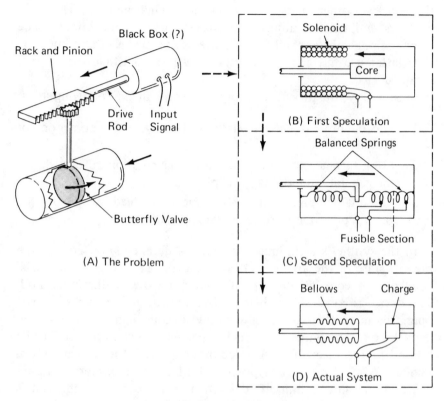

Fig. 4-7. The synthesis method.

creates a magnetic field which pulls the core into the solenoid. Previous experience with solenoids indicates that the method is workable but would require relatively large currents. He now considers some form of stored energy. A charged condenser might be made to "dump" through the solenoid and thus produce a high thrust for a short period. He then thinks of other storage methods including stretched springs. This second possibility, using two balanced, stretched springs, is shown in Fig. 4-7C. The signal wires are welded to one spring. An application of current fuses the latter and allows the drive rod to be pulled over by the second spring. This makes the apparatus non-resettable, but the nature of the application would justify this. The actual device is now dissected and found to contain a bellows and a small charge of explosive. Setting off the latter increases pressure on one side of the bellows and causes the valve to open. The original intention has obviously been to open the valve very rapidly, but some delay time is inherent in the transfer of explosive energy to the gas in the cylinder and in the inertia of the parts. The spring concept might be just as fast-acting if the fusible portion could be made small enough to allow very rapid heating. Further consideration shows that only one spring and a thin, fusible "restrainer" might do the job. The spring method might also be less expensive than the presently used scheme, in view of its simplicity.

Group Methods

All the creativity enhancement techniques described above can be used by groups as well as by individuals. The independent inventor often does not have the advantage of conferring regularly with a group to discuss his particular project. The growing number of independent inventors, however, has led to the formation of societies and organizations in various cities for the specific purpose of mutual aid.

The most well known of group methods is the "brainstorming" technique. The method is that of listing and combination. The procedure was originated and made popular by Alexander Osborn, psychology professor and industrial consultant in creativity. The ideal group, as proposed by Osborn, consists of 6 to 12 people meeting for periods of one hour. The description of the problem is sent to each member some time before the session. Each person is encouraged to present as many ideas as possible relating to the solution of the problem or its clarification. The entire session is tape recorded and copies of all illustrations drawn on paper or on a blackboard are retained. One member of the group acts as its chairman to coordinate the discussion. No criticism of any idea is permitted. When the session is over, the ideas are extracted from the tape and listed. Several days later, the chairman, the group itself, another individual, or another group evaluates the ideas, reducing the list to those appearing most valuable.

The question arises as to what are the ideal backgrounds for the people chosen to become members of a brainstorming committee. Should the committee consist only of inventors and technical people?

It has been found that a variation of points of view is extremely valuable. If one of the members is a salesman, another a lawyer, a third an artist, etc., it is often possible to obtain a wider and more imaginative range of ideas than if the group is professionally homogeneous.

Other variations of the brainstorming technique have been devised. The most well known of these is one developed by William J.J. Gordon of the Arthur D. Little Company, a research organization of Cambridge, Mass. This firm will invent a product to order for its clients. The Gordon technique is similar in many respects to brainstorming except that only the chairman is familiar with the exact nature of the problem. The rest of the group is asked to discuss a general principle (which is, of course, related to the problem at hand). If the chairman judges that the conversation is becoming irrelevant, he steers it back by revealing a little more of the true problem. The theory behind keeping the contributors practically in the dark is that they will not "center" too early on a particular thought pattern and will retain flexibility longer. As an example of the approach taken, let us assume that the problem is to develop a fog penetration system for use in automobiles. The chairman would ask the group a very general question: "How would you find your way around in a foreign country?" This starts the group thinking about various inquiry systems, how to get the most information using a limited vocabulary, the use of pantomime, etc. The next question would be "How do you look for something that is lost in the house?" Now the overall direction is towards methodical search methods. After the number of contributions to this subject begins to lag, the next question is: "How would an extremely nearsighted person be able to improve his travel?" This builds on the previously contributed material but now focuses attention on a limited range of vision and the need for position detection by means other than visible light. Finally, the chairman reveals the true extent and nature of the problem.

Some confirmed individualists have opposed group methods. Among the objections raised is the claim that the ideas obtained are superficial and will not represent more than simple combinations and permutations. John Steinbeck once wrote that no great idea ever came from more than one individual (working alone). This may be true but it is also true that many lesser ideas — and profitable ones —

did originate from the combined deliberations of several people. A useful effect found in brainstorming sessions is that of cross stimulation. The individual will often make contributions based on what someone else has said. If he were working alone, a certain percentage of his ideas may not have appeared because of the absence of cross stimulation.

Warm-Up Methods

"Mind stretching" by the solution of arbitrarily chosen problems is a good means of creativity enhancement and can be employed prior to working on an invention. In using this approach the individual or group sets up and solves a series of exercises that become progressively harder. The solutions can be preliminary or can be carried as far as desired. After several weeks of daily practice, the imagination will have been sufficiently stimulated to permit an effective attack on the basic problem.

One version of this method employs simple and multiple combinations. Table 4-4 is a list of 109 items chosen from an advertiser's catalog. Table 4-5 suggests specific improvements to be made in items from List I.

In practice, the method first requires the selection of two items at random from Table 4-4. The individual can choose items with which he is familiar. He then combines the two into a new device which is, to his knowledge, new and unique. Assume that he picks numbers 20 and 82, dog whistles and radios. A possible invention might be a tiny radio attached to a dog's collar and a transmitter carried in the owner's pocket. When the dog is to be called, the owner presses a button. The receiver emits a high-pitched sound in the collar. After the initial concept is written down, practical details can be worked out, e.g., the receiver and transmitter can both be built of integrated-circuit chips that can operate for very long periods on battery power. The transmitting frequency, the range of operation, possible interference with television, etc., are secondary considerations that can then be thought out.

After a number of these exercises are done, random choosing of items in Table 4-4 is no longer permitted. The individual must then take two "assigned" numbers, e.g., two consecutive items or two blindfold selections and combine these into an invention. The third step in these exercises is to choose three items and repeat the process.

Table 4-4. Items for Combination or Improvement.

1. Abrasives	38. Heaters	75. Oxygen masks
2. Absorbers, shock	39. Hedge clippers	76. Paint sprayers
3. Automobile polishes	40. Hose	77. Pants
4. Accordions	41. Ignition, pilot lights	78. Patches
5. Asthma relief	42. Inlays	79. Pens
6. Baby feeding	43. Indigestion aids	80. Phonographs
7. Bags, sleeping	44. Illustrating	81. Racing cars, toy
8. Bait	45. Iron cement	82. Radios
9. Banks, novelty	46. Jackets	83. Razors
10. Batteries	47. Jacks, auto	84. Reducers
11. Cabinets	48. Jewelry	85. Rings
12. Calculators	49. Juvenile furniture	86. Sandals
13. Cameras	50. Jig saws	87. Saws
14. Camp stoves	51. Kerosene weed killers	88. Shoe-shine apparatus
15. Carving tools	52. Key chains	89. Sinks
16. Desks	53. Kits, model	90. Smoking, stopping
17. Diamond saws	54. Knives	91. Tables
18. Dispensers	55. Kites	92. Tailoring
19. Diving gear	56. Labels	93. Tennis equipment
20. Dog whistles	57. Lamps	94. Testers
21. Electric drill	58. Lenses	95. Typewriters
22. Exercisers	59. Levels	96. Upholstery
23. Exhaust fans	60. Lawn sprinklers	97. Ukuleles
24. Exposure meter	61. Magnets	98. Ultrasonic devices
25. Eye protectors	62. Medicine, skin	99. Ultraviolet devices
26. False teeth	63. Marble, imitation	100. Underground pipes
27. Faucets	64. Massage	101. Vacuum jugs
28. Fenders, auto	65. Metal cutting	102. Vacuum cleaners
29. Figurines	66. Name plates	103. Vending machines
30. Filters	67. Number stamp	104. Waders, fishing
31. Gas burners	68. Nursery needs	105. Watches
32. Gem cutting	69. Novelties, tourist	106. Wagons
33. Glue	70. Nails	107. Washing machines
34. Golf equipment	71. Office supplies	108. Water heaters
35. Grinders	72. Oil burners	109. Xylophones
36. Hack saws	73. Openers	
37. Harmonicas	74. Outboard motors	

Table 4-5. Improvements.

1. More economical to produce
2. More economical to use
3. Easier to learn to use
4. Simpler
5. Adapt to home use
6. Change from "masculine" to "feminine" or vice versa
7. Employ alternate form of power
8. Remove a present limitation
9. Extend area of use
10. Improve public acceptance
11. Make larger
12. Make smaller
13. Change color
14. Change shape
15. Increase strength
16. Decrease strength

The fourth step requires that one item from List I be improved according to the instructions contained in one of the entries in List II.

One writer has referred to this process as "mental jogging." With practice it becomes much easier for the participant to apply his imagination to any inventive problem. When he now turns his attention to perfecting the desired invention, he finds the work much easier and his efforts more effective.

Pitfalls

The act of discovering an elegant solution to a problem imparts a sense of accomplishement and a feeling of high achievement. It is easy for the inventor to become so engrossed with the generation of new ideas that he does not wish to go on to what may appear as dull work. The preliminary technical feasibility of his solution satisfies him and he loses interest in the problem. As a result, some inventors leave many good concepts languishing inside notebooks and tucked away in desk drawers.

A cartoon which appeared a number of years ago shows the president of a television manufacturing company standing beside an empty, three-sided console containing openings for three picture tubes. He has called his engineers into a conference. "I figured out how to build a stereoscopic picture system. I will leave the details of the circuitry to you." The individual inventor, unfortunately, cannot turn over the "messy details" to engineering subordinates. He must test the idea for technical practicability, build various prototypes, compare with prior art, obtain a patent, redesign for mass production, set up tooling, explore markets, etc. Although he alone may not be this versatile, he must supervise the work of the various specialists he hires to perform each phase of the development for him. Most of the various techniques presented above for producing inventive ideas will also work well for problems encountered anywhere in the various stages of development, production and marketing — from preparing the first demonstration to determing what are the best media for advertising.

In addition to accumulating ideas without subsequent follow-up, the independent inventor is often guilty of insufficient preparation. The above described methods of creativity enhancement are based on recalling, combining, associating, simplifying, and speculating but can only operate on knowledge which is already present. Unless this knowledge is steadily increased, creativity enhancement techniques will produce a great number of "variations on a single theme" or a steadily decreasing number of original concepts. We see the re-invention each year of many old favorites because the inventors did not become well informed on what is available in the patent literature and on the market or what new techniques have been developed in each area of interest. Creativity is highest after a problem has been thoroughly studied and allowed to "digest."

5
Underlying Principles of Some Recent Developments

It makes an interesting and valuable exercise to examine an invention with the purpose of attempting to classify the characteristics that make that particular device unique and patentable. Did the inventor combine two or more simpler concepts to achieve a new result? Did he analyze the problem, prepare a list of possible solutions, and then methodically select the most promising for testing? Did he borrow from nature or did he change viewpoints to bring about a fresh approach?

We have collected a number of examples of recent inventions and will describe each in terms of prior art and the mode of thinking that appears to be dominant in each case.

Motorized Drum — Combination of Two Functions

A commonly used conveyor is comprised of two or more drums mounted in a ladderlike frame with a flat, endless belt stretched around the drums. A motor coupled to the front drum causes it to rotate and to drag the entire belt over the drum system. These machines are used to carry loose material, packages, subassemblies, etc., over various distances.

The motor drive is generally placed on one side of or under the conveyor frame, and power is transmitted to the front drum by means of belts, flexible drives or chains. Aside from the vulnerability of this arrangement to accidental damage or to clogging by material falling from the belt, much space is occupied by the drive and its mandatory protective shielding. In the case of short conveyors, which are transported from job to job, this configuration makes for clumsy handling and often results in harm to equipment.

In a recently marketed invention, the drive motor and drum are combined in a clever way. An electric motor is mounted inside the drum with power being supplied through slip rings. A simplified diagram is shown in Fig. 5-1A. The stator of the motor is fixed to the drum interior. Because both ends of the rotor shaft are rigidly held in the conveyor frame, the stator and the drum turn when power is applied. The machine also employs an internal gear box (omitted from the drawing for clarity) to achieve lower drum speed and greater output torque.

This particular combination of motor and conveyor drum not only overcomes the original disadvantage of a separate motor and coupling, it also achieves greater utility than a mere aggregation of parts would afford. In the present invention the dust seals normally used to protect drum bearings will now also prevent the entrance of dust and moisture into the motor, an important consideration in the reliable functioning of a conveyor. In addition, a quantity of oil can be maintained inside the drum for lubrication and cooling. This extends gear and bearing life and cools the motor. The relatively large area of the drum is ideal for rapid heat dissipation. Fig. 5-1B shows the motorized pulley in use.

Plug-In Switch — Problem Analysis Followed by Specific Solution

Switches in most electrical circuits are installed on a permanent basis (unlike light bulbs, for example, which are designed for easy removal). Replacement of a worn switch requires a relatively long interruption of service. In domestic applications this is not a serious problem because switch contacts last several years; the only inconvenience as a result of replacement is the need to reset a few clocks. In industrial circuits, on the other hand, failures are more frequent as a result of the heavy currents being switched and the accumulated "down-times" become expensive. There is need for a long-lasting switch or an easily replaceable

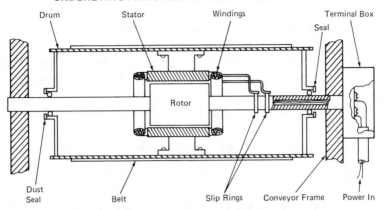

Fig. 5-1A. Motorized drum-simplified diagram.

Fig. 5-1B. Motorized drum in use. *(JOKI of America)*

one. The inventor could attack this problem by considering mechanisms for reducing arcing. Methods for suppressing arcs are known, but these involve expensive modifications. Another approach would be to list the characteristics of the problem and see what other alternatives are available.

A switch can conceptually be divided into two functional parts:

1. the part that electrically joins it to the rest of the circuit;

2. the part that makes and breaks the circuit.

In conventional switches part 1 is comprised of a nut and bolt arrangement for holding bare portions of the circuit wires, while part 2 embodies the make and break functions (contacts). Parts 1 and 2 are physically and electrically connected. If parts 1 and 2 could be made temporarily separable, it might be possible to replace part 2 quickly and easily. A switch built to these requirements would also need to be serviceable with the power on and provide safety for those doing the servicing.

The answer found by one manufacturer is shown in Figs 5-2A and 5-2B. The switch shown here is a roller-actuated type but is available in other configurations. A fixed module that contains the wire connections and an opening for conduit is attached to a mounting surface as with ordinary switches. An insulated, plug-in module contains the switch mechanism itself and is also provided with male prongs. These mate with corresponding receptacles in the first box. After plug-in, the second element is firmly secured to the first using the hold-down bolts shown. To replace the contacts it is only necessary to unscrew the bolts, substitute a new unit for the worn one, and tighten. It is unnecessary to turn off the electricity during replacement. The down-time is thus reduced to a few minutes.

One Piece Racquet — Listing of Shortcomings of the Present Art

An inventor attempting to improve an item of long-used equipment such as a tennis racquet might do well to first list its weak points and disadvantages. Fabrication cost would most certainly appear on such a list made up for a tennis racquet. A laminated wooden sheet is

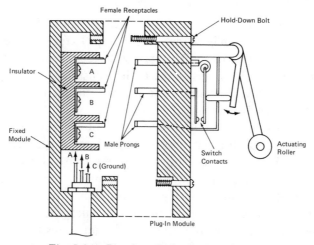

Fig. 5-2A. Plug-in switch construction.

Fig. 5-2B. Plug-in switch. *(Square D Company)*

first sawed into bars and these drilled to accommodate the strings. The bars are then steam bent into closed loops to form frames. Handle pieces are next glued on and wrapped with tape. Finally the string is threaded through the holes, kept under tension, knotted, and cut. Even with automated assembly the manufacturing process is complicated and inherently expensive.

Other items on the list of shortcomings of present day equipment would deal with performance: ordinary strings can rotate when encountering a ball; they also move with respect to those with which they are intermeshed. These motions dissipate some of the applied energy so that the ball's velocity suffers. In addition, the deflection of the string mesh at the impact area is conical; maximum bending of the strings thus occurs at the point of contact. This may or may not be the most efficient mode of operation.

A clever modification of racquet design that overcomes many of these difficulties is offered by a recent invention. The entire racquet — frame, strings and handle — is molded from one piece of plastic. The fabrication cost is thus considerably lowered. New problems associated with plastics such as cold flow, tensile strength limits, and shrinkage now need to be considered but can be handled by material selection and by proper design. The one-piece approach utilizes glass-filled nylon resin, a material whose shrinkage can be controlled by varying the glass content. It is found, however, that molding the strings at a 90-degree crossing angle with sharp corners results in the creation of stress raising points and a weakened structure. This problem is solved by the use of filleted crossover points (Fig. 5-3A). Approximately 1500 radii per racquet were needed, but these were incorporated in the mold and did not add greatly to the unit cost.

The conical impact area of the ordinary racquet can be retained in the molded version if desired. On the other hand, "springs" can be cast between the center string area and the frame. The center area thus acts like a trampoline to provide a more uniform drive from all parts of the racquet (Fig. 5-3B, page 105).

As often happens with an improvement, unforeseen possibilities became apparent when the concept is reduced to practice. In the present case it was found that the strings could be streamlined (Fig. 5-3C). This lowers air resistance when the racquet is swung and makes for higher ball

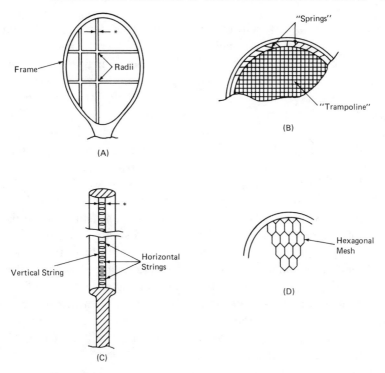

*String Dimensions Exaggerated for Clarity

Fig. 5-3. One-piece racquet. *(Saray Engineering Co.)*

velocity. Another possibility is the control of string tension. Unlike the present racquet where tension must be applied mechanically and the strings tied off, the one piece design benefits from the shrinkage of the plastic during molding. The amount of tension can be regulated by the amount of glass initially added to the resin. Still another potential improvement is the use of mesh angles other than 90 degrees. It would be a relatively simple matter, for example, to produce a hexagonal mesh if it were found that this design provided more efficient energy utilization.

Ultrasonic Ranging — Biological Modeling

The imitation of nature is one of the most powerful approaches in solving problems and developing inventions. Natural processes generally function so well for the purposes they serve that the experimenter would find it extremely difficult to better them. One example is found in the range-finding abilities of the ordinary bat. These animals can, in total darkness, fly through a room in which vertical wires are stretched at twelve-inch intervals. The bat traverses the room without as much as a single collision. This powerful faculty depends on a series of ultrasonic pulses that the bat emits. He times the interval between the generation of the sound and its reflection from obstacles and it thus able to judge his heading towards and distance from any obstacle and to adapt his flight accordingly. The emission of each pulse depends on the prior reception of a returning echo from the previous pulse. The frequency of the emitting mechanism is therefore a direct function of the animal's distance from the object.

Previous applications by man of echo timing have included such developments as sonar, crack finding in metal castings, and loudspeaker response testing. More recently a lens-setting range finder for cameras has been invented and marketed. This application more nearly duplicates the process found in nature inasmuch as the purpose is range finding in air over short distances and the use of a signal for controlling a mechanism.

The general principle of the new camera rangefinder is shown in Fig. 5-4A. The transducer, a stretched metallic diaphragm, is used both as a sound generator and as a condenser microphone. A short electrical burst from a pulse generator is applied to the diaphragm, produces a sharp ultrasonic beep and also starts an electronic, clock-driven counter. The diaphragm then remains inactive while awaiting the arrival of an echo. When the echo reaches the diaphragm, it is reconverted to an electrical signal, amplified, and then used to stop the counter. A digital-to-analog converter produces a voltage the magnitude of which is proportional to the number of milliseconds required for the sound pulse's round trip. This voltage is next compared to the previous input (which is defined by the position of a lens-drive motor). If there is a difference, the motor will turn to update the lens setting and will also create a new comparison voltage. If the range has not changed since the previous echo and return, the motor will not change the focus.

The time relation between the emitted beep and the returning echo is illustrated in Fig. 5-4B. The ultrasonic beep lasts for one millisecond. The round trip echo time varies from 3.6 milliseconds

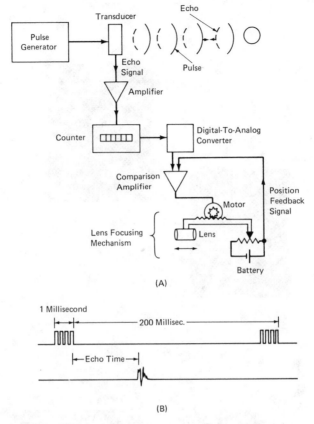

Fig. 5-4. Ultrasonic ranging system. *(Polaroid Corp.)*

for a target distance of 2 feet to 67.5 milliseconds for a distance of 38 feet.

Figure 5-4C, page 108, shows a kit offered by the manufacturer for use by inventors and designers in applying the principle to uses other than cameras. The circular objects are the stretched diaphragm assemblies used as transducers.

Roto-Jet ® Pump — Function Reversal

A powerful inventing technique mentioned in Chapter IV involves the changing of viewpoints. In one version of this method, the inventor imagines himself standing on one component of an existing

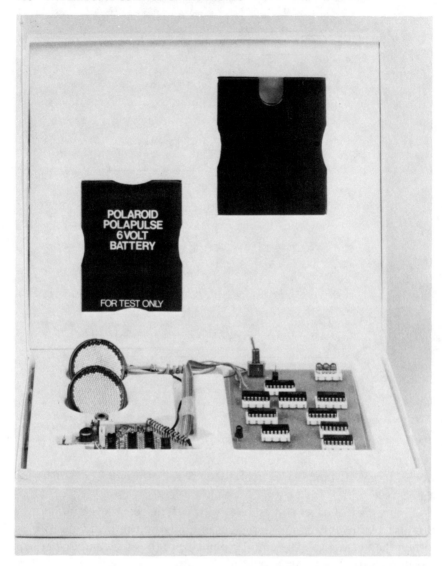

Fig. 5-4C. Ultrasonic ranging kit. *(Polaroid Corp.)*

machine as it operates, watching the motion of the other parts and noting the moment-to-moment relationships between his base point and the rest of the machine. He then moves to another component and repeats the observations. It is often found that an imaginative process of making the "tail wag the dog" suggests a substantial improvement.

A good example of function reversal is provided by a new type of rotary pump. Most machines of this kind use a rotor attached to a motor-driven shaft; the rotor is contained in a stationary, cylindrical housing, as shown in Fig. 5-5A. The pump is first primed, i.e., the housing is filled with the liquid to be pumped. When the motor is started, the rotor causes the liquid to be thrown by centrifugal force to the inner periphery of the housing. The liquid is then moved along by the rotor blades to a tangentially positioned outlet opening. This creates a partial vacuum at the center of the pump. More liquid then enters through the suction inlet pipe. This pump is characterized by

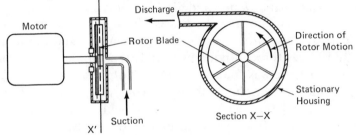

(A) Conventional Rotary Pump

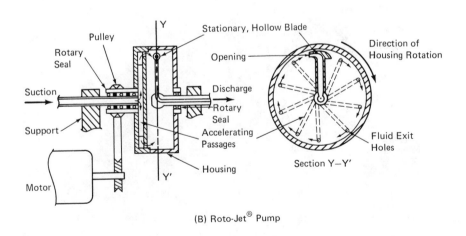

(B) Roto-Jet® Pump

Fig. 5-5. Rotary pumps.

high capacity at low pressure. The new pump, on the other hand, reverses the motional relationship between the housing and the rotor. The housing turns while the "rotor" is stationary. In this pump the rotor takes the form of a single, hollow blade open at both ends (Figure 5-5B, page 109). One opening is turned to face in the direction of rotation of the housing; the other opening connects to a discharge line. In operation, the liquid being pumped is driven by centrifugal force to the inner periphery of the rotating housing and is dragged along by friction between liquid and metal. The liquid now impinges on the up-stream opening in the blade, passes through the blade, and exits through the second opening to the discharge. Rotary seals permit the housing to turn while maintaining hydraulic connection to the stationary suction and discharge lines.

The housing contains radial passages through which the incoming liquid passes; these produce centrifugal "spin up" and acceleration. The liquid leaves through holes just inside the periphery of the housing (Fig. 5-5C). It was found that the use of passages produced better acceleration of the fluid and higher overall efficiency. The high velocity of liquid through the passages quickly removes air. As a result, the jet pump is self priming, another of its advantages.

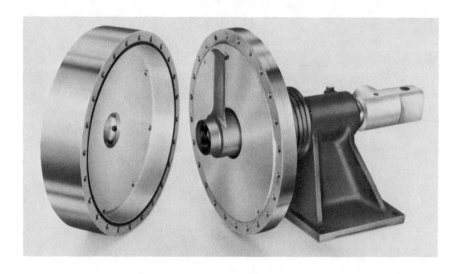

Fig. 5-5C. Roto-jet® pump dissembled. *(KOBE, Inc.)*

Push Pedal Bicycle — Listing, Analysis, Association

In the invention to be described, the listing of problem characteristics enabled the inventor to obtain a clear definition of what was needed. Once the objective was clearly delineated, the inventor surrounded himself with much thought-inducing material such as photographs and sketches of prior art and drawings of potentially useful mechanisms. Impressions and associations then led to the evolution of the idea.

The general problem was to devise an apparatus to minimize automobile use in congested areas. Two presently available methods for doing this are:

1. making use of urban buses, trolleys or subways;

2. riding a bicycle so that the parking problem is minimized.

Method 1 is suited for intermediate to long range travel but is inconvenient for subsequent short trips that require additional waiting, transfers, etc. Method 2 has limited range. To combine the advantages of both methods, the inventor decided to develop an easily folded bicycle that could be transported on public vehicles to the general area desired. Unfolding the bicycle would then permit the individual to travel between points within a one- or two-mile radius. The inventor initially studied existing types of folding bikes and drive mechanisms. He posted drawings of the various models on the walls and ceiling of his home so that he was constantly being reminded of the task. He also took several trips through a science museum in which many gear mechanisms were displayed. As a result of this high degree of immersion, he was able to devise a nonconventional drive that is easy to fold and makes for a lightweight construction.

The drive is shown schematically in Fig. 5-6A. Cables from pivoted pedals replace the usual chain drive. Each cable extends from its pedal through an idler pulley, around a special, one-way pulley on the back wheel, and terminates in a spring mounted to the bike's frame. When each pedal is depressed through its permissible arc, its cable acts on its corresponding one-way pulley to make the rear wheel turn. When the rider lets up on the pedal, the one-way pulley disengages from the wheel and permits the spring to return the pedal to its up position. Unlike the conventional bike, each pedal operates independently; both pedals return to the same position at

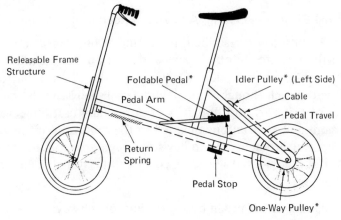

Releasable Frame Structure

Foldable Pedal*

Idler Pulley* (Left Side)

Pedal Arm

Cable

Pedal Travel

Return Spring

Pedal Stop

One-Way Pulley*

Steve Titcomb, Inventor

*Right-Side Pedal, Idler Pulley & One-Way Pulley are Identical

Fig. 5-6A. Foldable bike. *(Steve Titcomb, Inventor)*

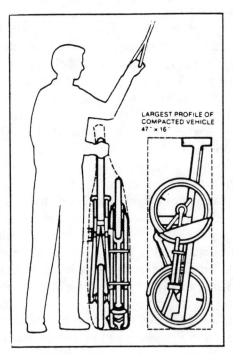

LARGEST PROFILE OF COMPACTED VEHICLE 47˝ x 16˝

Fig. 5-6B. Bike folded for carrying.

rest. The pedals can be folded out of the way when the bike is to be dissembled for carrying so that the transmission occupies minimum space, as shown in Fig. 5-6B. When folded, the bike fits into a volume of 45 by 12 by 12 inches. Total weight is twenty pounds.

Holography and Human Memory — Adaption and Cross-Fertilization

On some occasions biological modeling will work in reverse, i.e., a man-devised principle can be used to understand or improve a naturally occurring mechanism. It was found, for example, that laser interference photography displayed some unusual characteristics. In this process the film is exposed to two light sources: radiation coming directly from the laser and indirectly from reflection off the subject. The combined exposure produces an optical interference pattern that is recorded on the film. When the developed film is illuminated by laser and projected onto a screen, an image of the subject is obtained. The remarkable difference between this image and that produced in ordinary photography is that information for the visual details are distributed over the entire area of the developed film (the hologram). It is therefore possible to break off a small portion of the film, illuminate it by laser, and obtain the entire image. In ordinary slides each part of the image is geometrically related to a corresponding portion of film.

Another interesting feature of the hologram is its three-dimensional effect. As the observer moves his head with respect to the film, he can see details of semihidden surfaces of the subject. The interference pattern produced as a result of the arrival of light from the furthest portions of the subject permits accurate recording of what would normally be considered as "side" or "top" view.

The adaption of the holography principle to an explanation of human memory can be considered one of the outstanding concepts of recent times. No satisfactory analogy between computer memory and the brain can be drawn. Human memory has so small a search and recovery time that its operating principle must be based on other than that used in serial storage. A significant characteristic of the brain's memory mechanism is its storage of all information over a large area so that item recall is possible even when considerable tissue has been destroyed. If the key to the brain's memory mechanism has some parallel in holography, it may be possible to discover the entire operating mechanism and to make great strides in medicine and education. Further cross-fertilization between biological and optical discoveries might also lead to improved electronic memories for better computer and control system design.

6
Planning the Experiment

Most ideas work well on paper. When put to the test, though, many develop functional "diseases" which may or may not be curable. To reveal these shortcomings, the investigator turns to his most valuable proving tool – the controlled experiment.

When the average inventor comes to the point of trying out his concepts, he is often faced with a number of unexpected difficulties. Unusual conditions may call for more than ordinary mechanical skill. His experimental apparatus may prove too crude and the results from it may be inaccurate. He may overlook certain factors, which leads to erroneous conclusions; the latter are then used to modify the invention until a point of complete perplexity is reached. He may, on the other hand, "overbuild" and invest more time and money than is necessary to establish the feasibility (or non-feasibility) of an idea.

The professional scientist is not completely immune from these problems but has been trained in a number of techniques for the planning and carrying out of his experimental procedures. The use of these techniques tends to avoid non-fruitful experimentation and to minimize poor and misleading results. A number of these methods will be discussed in this chapter.

General Principles of Experiment Planning

An experiment serves several purposes. First, it tests a theory which may have occurred to the inventor in the form of "Wouldn't it be nice if. . .?" The results are a general confirmation or denial of the theory. A second purpose of an experiment is to collect data from which calculations can be made and conclusions drawn. A third purpose is to help develop a viable model. Experimental measurements are often directly translatable into working model dimensions. A fourth purpose is to establish a reduction to practice. By actually constructing and testing a working device, the inventor rigorously meets one of the requirements for patentability. Let us illustrate these four purposes by a hypothetical example. A certain inventor is seeking a new type of light source. The light-emitting diode is appealing because it operates almost completely without heat emission. The diodes presently on the market are tiny, both in physical size and in light output. The inventor visualizes increasing the size of the semiconductor and thus producing more light. An experiment with a larger semiconductor chip does seem to bear out this reasoning, but the gain in output is not in proportion to the increased size. Further enlargement of the diode actually decreases the light output per watt of input current. Experiment has now served to evaluate the concept in a general way. He then conducts a series of tests to find the optimum chip size for his purpose — maximum light, minimum current. This is the data-gathering phase of the work. He now realizes that he must combine a large number of separate chips into each light "bulb." He next experiments with means for etching through a mesh and obtaining an array of chips from a single slab of semiconductor. He is now developing a practical concept. He finally achieves a workable process for preparing a "glow surface" containing many individual diodes. This is the reduction to practice step.

A properly designed experiment will have the following characteristics: it will test for the effects of one phenomenon or variable and will hold all other variables as constant as possible. When repeated, it will give essentially the same results over and over; it will be as simple as possible consistent with the results sought; it will produce, whenever possible, numerical data which can be evaluated by others; and it will handle "blanks" (internal standards, references, etc.) by which the functioning of the apparatus can be continually checked. To help the inventor achieve these aims, we have drawn up a general list of recommended steps. A number of examples have been included to demonstrate the use of this procedure.

Steps in Experiment Layout

1. Write down the purpose of the experiment. If, at a later time, the work seems to stray down a side path, this statement will remind you of your primary aim.

2. If the goal is complex, plan on two or more experimental programs. In one type of problem, several experimental programs can be run in parallel; one is independent of the results of the others. In another type of problem, the start of one set of experiments depends on the successful outcome of its predecessor.

3. Consider functional concepts of the experiment first. A convenient way to do this is by a block diagram showing each step of the work. Use symbolic notation to indicate possible outcomes. The symbols employed in computer programming are very useful for these diagrams.

4. Simplify the functional diagram, if possible.

5. Design the practical experiment. Convert symbols into the actual hardware required.

6. Rework step 5, if possible, to simplify, reduce size, decrease the number of parts. Make sure functionality is not impaired during rework. If specially fabricated pieces are indicated, try to find manufactured items which can be substituted.

7. Add means for taking measurements. Remember the importance of numerical data.

8. Design a data-handling scheme. How will the various numbers be treated so as to permit the drawing of unbiased conclusions?

9. Give preliminary thought to a "second generation" set of experiments. If certain results are obtained, a new set of experiments may be required.

10. Set up and run the experiment.

The order of the steps is flexible; some interchanging may be desirable in individual cases. In some problems certain steps may not be necessary.

As an example of the use of this procedure, we shall help an inventor plan an experiment to evaluate the following concept: ship lifeboats are often equipped with emergency radio transmitters with which occupants can call nearby vessels. The range of these transmitters is limited by the length of antenna available. If a high enough antenna could be raised for even a short time, the area covered by the signal would be great and the chances of being heard by a passing ship measurably increased. The inventor wishes to use the sea itself as the antenna; his idea is to create a long, upward jet of salt water to which the transmitter would be electrically connected. He proposes to do this by some pumping scheme which would discharge water high into the air in the manner of a fire hose. The inventor has reason to believe that a column of water would be a much more reliable antenna than a thin wire held aloft by a balloon. Previous balloon systems were difficult to inflate and the thin wire required would often become snarled during storage.

We wish now to design a set of experiments to study the above concept. We will use the following symbols in our block diagram:

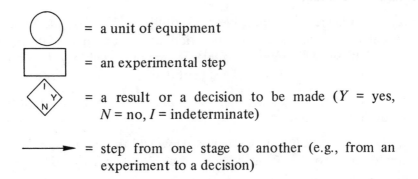

○   = a unit of equipment

▭   = an experimental step

◇ (I, Y, N)   = a result or a decision to be made (Y = yes, N = no, I = indeterminate)

⟶   = step from one stage to another (e.g., from an experiment to a decision)

Using these symbols, we draw up the initial plan shown in Fig. 6-1. The object of the experiment will be to put a quantity of salt water through a pressurizing device (perhaps a pump), then through a jet-forming device such as a nozzle. A radio will be connected to the latter. If an antenna is suitable for receiving, it can be tuned to act as a good transmitting antenna as well. All possible outcomes are shown by the diamond in Fig. 6-1. Reception will or will not be

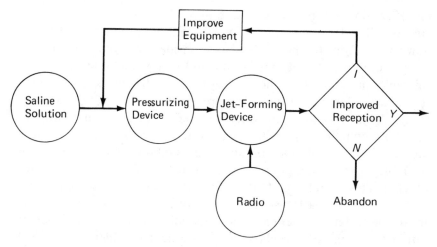

Fig. 6-1. Antenna experiment – initial plan.

improved or the results will be indeterminate. If indeterminate, we will modify the experiment and try again. If definitely negative, we will abandon the idea.

We now consider what data should be gathered and what action should be taken for all the possibilities shown in the initial plan. The following items would appear to be of interest:

1. Effect of height on reception. How to measure jet height?
2. Effect of jet break-up. Will this cause static or noise?
3. Can the concept be used in a high wind or storm?
4. How can reception effectiveness be measured?
5. What kind of pump shall we use?
6. How to couple the radio to the jet?

We next begin to visualize practical aspects of the problem. A small pump would take up water through a suction line lowered into the sea. The water would be ejected through a nozzle. Two immediate objections come to mind: the jet and the main body of the sea might be electrically continuous so that the antenna would be "grounded;" a vertical jet would return directly down into the lifeboat and possibly create problems. To avoid these objections, two modifications will be made: the pump will draw water from a previously filled tank and the jet will be aimed at an angle away and down wind from the boat. We next consider what our experimental setup should be. A motorized pump and tank would represent a considerable investment. The concept could be checked out by a stationary column of water. The apparatus of Fig. 6-2 is one possibility. A jar containing salt water is closed off by a three hole stopper. One hole contains a glass tube of arbitrary length — say 15 feet of tubing. The second hole contains a glass elbow which is connected to a bicycle-type air pump. The third hole contains a brass rod which makes contact with the salt water and is connected to the antenna lead-in of a transistor radio. By pumping air into the space above the water, we can create a water column of any desired length within the range of the glass tube. We next consider how we will

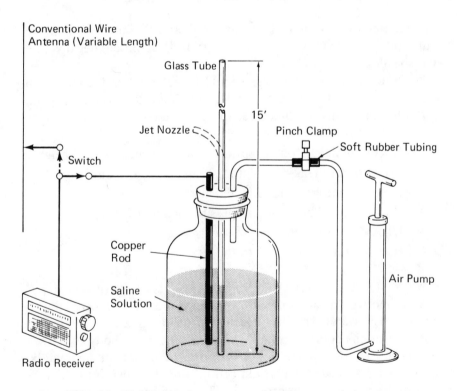

Fig. 6-2. Experimental arrangement for antenna experiments.

make a measurement. The radio chosen shall be of a type having numbers marked on the volume control dial (evenly spaced from 1 to 10). A few experiments with the radio show that it is possible to obtain a number which represents the bare limit of audibility of a particular station. This number would be expected to vary as the length of the water antenna is altered. Another measurement technique will be to substitute an equivalent wire antenna. At constant setting of the volume control we shall switch the receiver rapidly from the water unit to varying lengths of a copper wire mounted vertically. The "wire equivalent" in feet to give equal loudness would be another measure of performance. Our planning points up a source of possible error. Personal judgment is involved in determining minimum audibility and equal loudness. Reference to any text on radio indicates that there is a much better way to do this. All radios are equipped with an automatic volume control system which maintains constant loudness by feeding back varying control voltages depending on signal variations at the antenna. A voltmeter used to monitor this feedback gives an excellent quantitative indication of signal strength. We will use the voltmeter method in preference to the others from this point on. It is now assumed that the stationary water column gives promising results. We will next plan on cutting and bending the glass tubing as shown by the dotted lines in Fig. 6-2. This will form a free standing jet when pressure is applied. To maintain a steady jet, we must add another source of air. An inflated inner tube or large balloon will provide a temporary air supply. It will be desirable at this time to consider a small air compressor; it can be used for later experiments. Our data will consist of a tabulation of signal strength as a function of jet height. The saline solution will be made up to approximate the composition of sea water but we may wish to check other concentrations as a side study. We assume positive results and plan further and more elaborate experiments. A radio signal generator can now be incorporated into the experiment. This unit is placed at some distance from the apparatus and tuned to produce an audio-modulated signal in our receiver (at some unused frequency on the dial). The signal strength, as read by the voltmeter, will no longer vary with voice or music modulation and will be a truer measure of the effectiveness of the water antenna. The next step in perfecting the invention is the interchanging of the generator and the receiver so that the water antenna now acts for the first time in a transmitting mode.

It is assumed during planning that encouraging results are obtained at every step. The plan should also incorporate a contingency section which defines steps to be taken should negative results be obtained at any point. Other questions which may occur during the planning are:

1. Can the system be redesigned so that sea water is continuously drawn from suction but still retain electrical isolation of the jet?

2. If batteries are used to power the pump, how much transmitting time does this allow? Would it be possible to provide a system for manual pumping?

A little thought shows that any device which would prevent a solid column of water being formed between suction and discharge might serve as an isolator. The system shown in Fig. 6-3 would meet this requirement. A hand crank could drive both the pump and an electric generator for the transmitter. This would alleviate the power problem.

As the design of the experiment has evolved, the original concept has become more refined and has developed into a practical system. Planning could be continued from this point to include prototype construction and testing at sea but we have gone sufficiently far to demonstrate the method. The experimental plan is summarized in Table 6-1.

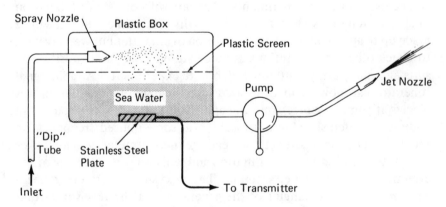

Fig. 6-3. Water antenna.

TABLE 6-1. WATER ANTENNA — TESTING OF THE CONCEPT.

Experiment Number	Apparatus and Test Conditions	Test	Results Yes	Results No	Indeter	Data
1	Bottle + saline solution, stationary column, radio using calibrated volume control	Overall feasibility	Proceed to next step	Repeat, if still no, abandon	Revamp and repeat	Yes, No Indeter
2	Bottle + saline solution + jet	Effect of free flowing jet	"	"	"	"
3	Bottle + saline solution + radio using monitor on automatic volume control system + jet	Effect of column height	—	—	—	Graph of column height vs. signal strength
4	Bottle + saline solution + signal generator placed some distance away	Effect of column height (more accurate and quantitative results)	—	—	—	"
5	Bottle + saline solution + signal generator coupled to jet (receiver + voltmeter remotely placed)	Effectiveness as transmitting antenna	Proceed to next step	Recheck experiment	Revamp and repeat	"
6	Bottle + saline solution + pump (+ conditions of test number 5)	"	"	"	"	"
7	"	Effect of salt concentration	—	—	—	Graph of signal strength vs. salt concentration
8	Tests at sea					

Simplification

In the previous example, we simplified the experiment by first using a stationary water column and a radio receiver rather than a transmitter. Simplification is a highly desirable procedure in experimental work because it permits more rapid evaluation. A heavy investment in time and material made in following a non-profitable idea is especially wasteful for the independent inventor because he is diverted from the exploration of other ideas. There is, however, another aspect of the simplified experiment. Extreme simplification can be wasteful because it leads to "jury rigged" or "quick and dirty" methods. Important phenomena may go undetected because the apparatus is not sensitive enough. The real skill of an experimenter lies in his being able to simplify to an extent sufficient to permit rapid study, but stopping short of an inadequate setup.

To illustrate the benefits of simplification, we will consider now the evaluation of an ore dressing concept. The inventor, in this case, seeks to eliminate the difficulties associated with the screening of various sizes of a certain raw ore. A local chemical plant passes the ore as received over a set of inclined, vibrating screens. Each screen is of smaller mesh than the previous one so that a series of graded powder sizes is obtained. The various sizes are then treated by different processes before being incorporated into the company's products. Problems occur because screening is slow and the screens frequently become blocked with fine powder. The inventor visualizes a trajectory process in which advantage is taken of the variation of flight path length with particle diameter. The inventor has hurled a mixture of stones of various sizes horizontally and noted that the larger pieces travel further than the smaller. Although a small particle encounters less air resistance than a large one, the greater mass of the latter provides higher momentum and a lower deceleration rate. As a result, the finer particles are decelerated rapidly while the larger ones undergo a longer flight. The inventor now wishes to determine whether there is a useful gradation of particle sizes while a mixture is in flight. His initial concept is shown in Fig. 6-4. Ore is introduced from a hopper onto a rapidly moving conveyor belt and is projected horizontally from the end of the conveyor. A series of bins placed at the end would collect particles of gradually increasing diameter. The

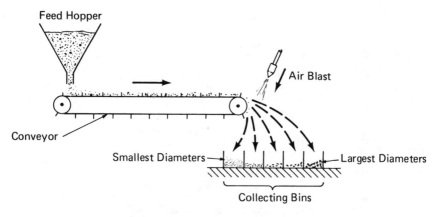

Fig. 6-4. Ore sizing concept.

finest material in the ore has the grain size of flour and would not be expected to travel very far after leaving the conveyor. Some of it will, however, adhere to the larger particles. To overcome this, the inventor visualizes an air nozzle at the end of the conveyor to forcibly remove this fine material and direct it towards the first bins. The inventor must now design an experiment which will evaluate the concept. After some study, he sees three areas of simplification:

1. The size of the experiment. Miniaturization will save time and money.

2. The driving system. He will use gravity to achieve particle motion.

3. Ore simulation. He will use a synthesized mixture to permit a more definitive initial study.

The proposed apparatus is shown in Fig. 6-5. A wooden frame is used to support a curved chute made of sheet metal. The latter is shaped so that vertical acceleration of the particles is transformed into horizontal velocity. Rectangular plastic trays are arranged to serve as collecting bins. A plastic funnel is cut to permit free flow of powdered material and mounted on a laboratory ring stand near the top of the runway. Plastic end walls prevent side spillage. The height of the funnel above the apparatus will determine the particle speed achieved.

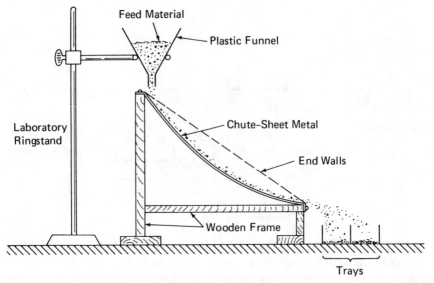

Fig. 6-5. Ore sizing — simplified experiment.

The inventor plans on running his initial tests with binary mixtures of spherical particles. A surprisingly wide variety of materials are manufactured in spherical form. He can use lead, steel, ceramic, carbon, or glass spheres in any diameter he wishes. He can also obtain some of these materials in hollow form so that a wide range of ore densities can be simulated.

The proposed ore sizing method is a form of separation process. Since the latter is a frequently used industrial technique, it will be helpful at this point if the inventor studies a few of its characteristics. In its simplest form, a separation process consists of passing a combination of substances, A and B (Fig. 6-6), through a separation device. The combination may be a mixture, solution, suspension, etc. It is desired in this general case to obtain a quantity of material enriched in component A. The starting composition is 10% A and 90% B. The hypothetical figures given in Fig. 6-6 show that 10 lb. of enriched material containing 92% A are obtained from this particular process from every 100 lb. of feed. The "waste" stream contains less than 1% of component A and amounts to 90 lb./100 lb. of feed. The efficiency or "figure of merit" of this process can be

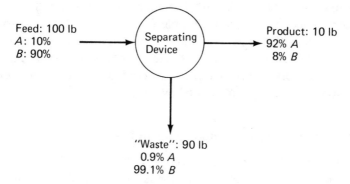

Fig. 6-6. The separation process.

calculated in a number of ways. The simplest is to express the efficiency as the ratio of the actual enrichment to the maximum enrichment possible. If the process were perfect, the product stream would contain 10 lb. of A and 0 lb. of B. The process of Fig. 6-6 would therefore have an efficiency of 92% (9.2/10 x 100). If further refinement is desired, it is often possible to recycle the waste stream through the apparatus or to pass the enriched stream through another apparatus. With these concepts in mind, we return now to the ore sizing experiment. The inventor decides to start with a 50% mixture of two sizes of glass spheres. After passing this mixture through the test device (Fig. 6-5), he weighs the contents of each tray and then carefully separates each of the contents into larger and smaller sizes. He can now calculate the efficiency of separation. A number of repeated tests will tell him how reproducible the experiment is and will permit him to derive "confidence limits" in his data. If the separation appears promising, he can then plan an extended series of experiments. The following would appear to be a logical choice:

1. Varying the percentage of the larger spheres.
2. Use of differing size pairs. Vary the larger size or the size difference.
3. Employ mixtures of three or more sizes.
4. Use actual ore samples.
5. Add a source of compressed air and study the effects with very fine spheres and with ore.

His experimental work with the simplified model will permit the inventor to accumulate considerable experience and data. From this he can modify the apparatus, adjust working parameters, and then scale up to larger and more elaborate experiments employing motor driven conveyors. Factors such as conveyor speed, air velocity, projection angle, etc., can then be evaluated.

Chemical Processes and Compositions of Matter

The planning of an experimental program to develop a chemical process or a composition of matter is similar to that employed for equipment development. The steps of writing down one's aims, composing block diagrams, etc., are again followed. Simplification, data handling, and experiment modification play a similar role in this kind of development. Examples will illustrate some techniques for planning this type of experiment.

In the first case to be considered, the following background information is available: the inventor wants to perfect a stronger plaster of Paris for use in decorative casting. Plaster of Paris is made by roasting gypsum powder in a slowly turning steel drum to drive off chemically bound "water of hydration" in each crystal. This is a fairly rapid process which results in the forming of very fine needles of plaster. When water is mixed with the plaster, the needles again form chemical bonds with water but now interlock to produce a tight mass. Unfortunately, cast plaster is brittle and soft; these characteristics make it undesirable for casting large decorative objects. The inventor finds a patented process in which the gypsum is dehydrated under water in a heated pressure vessel. Under high pressure it is possible to raise the internal temperature above $212°F$ without boiling. Chemically bound water is driven from the gypsum despite the fact that the reaction is occurring in an aqueous medium. When dehydration is complete, the vessel is rapidly cooled and opened; the plaster-water mixture is dumped through screens to separate the major portion of the water. The screens are then washed with alcohol, which removes the last traces of moisture. After the alcohol has evaporated, the dehydrated crystals are found to be much larger than those normally obtained by roasting. When the plaster of Paris prepared in this way is cast, it is several times stronger and harder than the ordinary material. There is a correlation between crystal size and the quality of the product; the larger the crystal, the better the plaster. The inventor theorizes that the uniform heating and the relatively slow removal of chemically bound water in the pressure vessel must induce the slow growth of large crystals. In the roasting process, on the other hand, the heating rate is high so that steam is formed inside each crystal and thus disintegrates it. The inventor believes he can add heat gently by a radio frequency technique. The added cost of the equipment and the electric power will be offset by savings to be achieved from simpler processing requirements and by eliminating the need for the alcohol drying step. His objective then is to evaluate radio frequency heating as a processing means in the preparation of high quality plaster of Paris. He draws up the initial plan shown in Fig. 6-7.

The inventor now prepares a list of the points to be considered:

1. How will the preliminary experiments be implemented? Will a kitchen type of microwave oven be sufficient? Will the gypsum "absorb" enough of the radio frequency energy to raise its temperature sufficiently? Shall the gypsum be confined

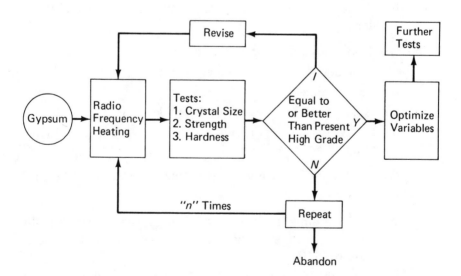

Fig. 6-7. Process for the production of plaster of Paris.

in a closed and insulated container? What material for the latter?

2. How will the temperature reached by the powder be measured? Dumping the heated powder in a measured amount of a nonreactive liquid and measuring the rise in temperature with a thermometer might be one method. What liquid to use?

3. What tests will be performed?

   a. Crystal size. Use a microscope with a measuring eyepiece.

   b. Strength. This can be studied by casting bricks of plaster and measuring how many pounds of force are required to pull them apart and to crush them. Look up standard methods of testing plaster for strength. Shall test equipment be rented or shall a commercial test laboratory be used?

   c. Hardness. Check resistance to scratching. Look up standard methods.

4. What materials will be used for comparison?

   a. Compare with commerical plaster.

   b. Compare with "extra grade" material made by pressure dehydration.

5. What variables should be studied?

   a. Temperature in radio frequency oven.

   b. Heating time.

   c. Is slow cooling necessary or desirable?

   d. Purity of starting material. What gypsum purities are available?

6. How will data be handled? How many tests should be run?

7. Special studies.

   a. Can anything be added to the gypsum (e.g., metallic powder) to improve heating characteristics in the microwave oven?

   b. Can radio frequency heating be used in similar processes – e.g., in the preparation of Portland cement?

Using the library, patents, and inquiries to various testing companies, the inventor obtains as many answers as possible. He can then prepare a table of the experiments needed to evaluate the concept.

A somewhat different problem is encountered when the invention involves pharmaceutical products. In the case we will now consider, the inventor wishes to perfect and market a lotion effective against poison ivy. He has had two experiences which led him in this direction:

1. A bad exposure during a camping trip. At the time, no available lotion was effective in decreasing the swelling.

2. The accidental spilling of aluminum acetate solution on his skin during a laboratory experiment. The chemical seemed to have a constricting or stringent effect, causing the skin to shrink at the area of contact. Subsequent experiments using this compound on swelling and irritation have shown very favorable results with no side effects.

He now believes that aluminum acetate may be effective in a poison ivy lotion. A study of the patent literature shows no prior art using this chemical. To set up an experimental program he first makes up the following question list:

1. Is a lotion the best form of treatment? Aluminum acetate is a powder and could be applied in this form. What are the advantages of dissolving it first? If dissolved, should the liquid be volatile and evaporate rapidly or should it be in the form of an ointment?

2. What other ingredients should be considered?
   a. Anti-bacterial agent in case skin is broken during encounter with the ivy. Perhaps alcohol will serve both as a solvent and as a disinfectant.
   b. Pain deadening additive. An ingredient such as benzocaine might be useful in this respect.
   c. Soap. Cleans the surface and emulsifies skin oil to make contact with the aluminum acetate more effective.

3. How to evaluate?

    a. Obtaining samples of poison ivy. What kinds are there? What other poisonous plants should be considered? How to prepare a "standard" irritant with which to paint subjects? Self protection during experiments.

    b. Experiments with animals. White mice, hamsters? Does furry area need to be shaved first? How can swelling and its subsidence on an animal be measured? Consultation with the biology department of a local university. All preliminary tests to be made with simple solutions of aluminum acetate in water (varied concentrations).

4. Test with human volunteers. Consult with federal and state authorities on legal aspects. Use of "placebos" — solutions containing no active ingredients — on some of the volunteers. This will permit evaluation of psychological effects since some individuals will improve spontaneously if they *think* they are receiving a beneficial treatment. Selection of volunteers: age groups, complexions, etc. Payment of volunteers.

5. Questions to be asked of volunteers:

    a. Describe previous allergies. May rule out certain individuals.

    b. Describe condition of the area painted with test solution — itching, redness, burning, swelling, no effect?

    c. Describe effect of the lotion: relief of itch, decrease of swelling, numbness?

    d. How long does it take lotion to act?

6. Handling of data. What percentage of volunteers report beneficial results?

7. Evaluation of additives. What is effect of these on the action of aluminum acetate?

A block diagram of the experimental approach is shown in Fig. 6-8.

In this type of experimentation, it will be noted that some of the test results will be qualitative and subjective. The investigator cannot

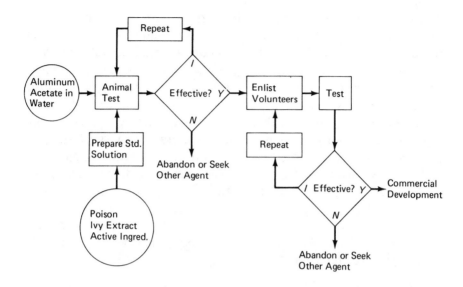

Fig. 6-8. Poison ivy lotion — experiment plan.

really determine "how much" or an unqualified "yes" or "no" in many of the tests. He must record an opinion as to the effectiveness of his lotion. The experiments may or may not show a high degree of repeatability, they may vary with individuals and with mental and physical states. In addition to these factors, the inventor will need to take into account the many variables introduced by concentration variations and the use of other ingredients (see the discussion of matrix methods, Chapter 4). When the development is to be patented, he must determine the optimum concentrations of each of the ingredients and state the ranges through which each can vary. His final composition may, for example, be specified as follows: 2 to 5% by weight of aluminum acetate, 1 to 3% benzocaine, 5 to 10% calcium stearate (emulsifying agent), 0.5 to 1% potassium permanganate (disinfectant), and the rest distilled water.

Two factors often enter into pharmaceutical experiments, and planning should take these into account. It is sometimes impractical or impossible to determine the actual composition of a useful formulation. The inventor combines weighed quantities of A and B in the presence of C (a catalyst), heats the mixture to temperature in range of T_1 to T_2, adds a certain weight of D, and allows the material to cool for h hours. The final product has certain useful properties and cannot be made in any other way. It is a complex compound which is very difficult to analyze. The approach used by the inventor involved the systematic variation of input quantities and conditions and the correlation of these variables with test data. In obtaining a patent, he would protect his right to the product by specifying the process − a legal procedure known as "product by process claiming."

The other characteristic of pharmaceutical inventions is the effect of dosage. The quantity used and the number of times it is used in a given period is as critical as its composition. It is sometimes found that unexpected and useful results exist at both ends of a dosage scale. A small dosage achieves one purpose; massive dosage results in an entirely different, but beneficial, result.

Overlooked Primary Factors

When initial experiments appear very promising, there is a temptation for an inventor to leap ahead in an effort to hasten success. At this point he sometimes casts aside a carefully drawn plan and omits a critical experiment. The need for adhering to an experiment plan is well illustrated by the fate of one inventor who had hoped to sell the government a large quantity of cold-weather uniforms. The inventor had carefully studied the problem of keeping a soldier warm in arctic temperatures.   Procurement regulations allowed normal winter clothing but did not permit body stoves, sewed-in electric coils, or other personal heating devices because fuel or batteries would need to be carried by the already heavily burdened service man. The inventor had determined that major heat loss under these conditions is through the breath. This occurs because extremely cold air is inhaled, heated by contact with lung tissue, and then exhaled. No heat is recovered.

The inventor devised a face mask with two ports, each containing a one-way valve. The ports alternatively opened and closed with every complete breath. The exhaled air was directed through a long, flexible heat exchange duct embedded in a special one-piece suit. During its travel through the duct, the air warmed the suit. Because of the lowered temperature differential between suit and skin, body heat loss was reduced.

The inventor built a working model and tested it in a refrigerated room. It appeared to function very well. He could sit in the room for several hours without discomfort. He grew enthusiastic and asked government representatives to witness the tests. As a result a considerable amount of optimism was generated. This led to the construction of a field model and a test in Alaska.

Another government requirement for cold-weather clothing is that a soldier clad in it must be able to run a short distance at low temperature while carrying a field pack and his rifle. In the test the inventor's suit failed miserably. The soldier fell exhausted after a few steps.

The inventor had not followed his original plan of field-testing the device himself. If he had, he would have discovered the invention's fatal shortcoming. The slightly higher pressure that the lungs were required to exert (so that each exhalation could traverse the heat exchange duct) made extreme demands on an already heavily burdened man. Human endurance was thus exceeded.

Summary

The planning of experiments can be thought of as a five-part procedure:

    a. determining what answers need to be found in order to solve a particular problem;

    b. devising a system which is capable of being interrogated;

    c. composing questions to apply to the system;

    d. modifying the system as necessary until all the questions can be and are answered;

    e. combining the answers into a useful and profitable result.

7

Apparatus Construction, Measurements, and Data Handling

Introduction

Some inventors have succeeded because they possessed an extraordinarily high degree of mechanical skill in addition to their innovative abilities. Others, less inclined to mechanical construction, have delegated the building of their apparatus to artisans over whom they kept close watch. Still others have used clever substitutions and have avoided the need for precisely made components. And finally, there have been those with great talent in the careful conducting of experiments. Even though they used relatively crude apparatus, they were able to obtain meaningful and significant results.

As an example of one extreme in elegantly-built apparatus, one need only consider the mass spectrograph built by physicist A.J. Dempster in 1918. A beam of ions (electrically charged atoms or groups of atoms) is introduced into an evacuated chamber where the ions are subjected to carefully controlled magnetic and electric fields. Various individual atoms in the beam are thus "sorted" out. The equipment is elegant and employs sophisticated control elements. The use of the mass spectrograph helped prove that the weight of an atom was greater than the sum of weights of its individual parts — the difference being, of course, what could be obtained in energy by fissioning. At the other extreme, we can inspect the relatively crude experimental setup used by the physical chemist A.J.P. Martin to develop the chromatograph. A visitor to Oxford University once requested to see the laboratory in which Martin perfected his invention. The visitor was conducted over wooden floors roughened by centuries of wear to a chemically stained bench on which were stacked a number of soiled glass tubes and a few simple electrical instruments. The highly versatile chromatograph developed from this humble equipment now plays an important role in such diverse fields as medicine, space exploration, pollution monitoring, and industrial process control.

It would thus appear from any study of past successes that wide deviations in working techniques for experimentation are possible. Whether their highest skill was in the designing, the building, the data taking, or the analysis stages, successful inventors seem to have had the following characteristics in common: persistence, great patience, and the knack of paying close attention to detail.

In this chapter, we will be concerned with the mechanics of experimental work. After a program has been planned, concepts simplified, and provision made for the collection of data, further development calls for the following tasks:

1. Preparing working sketches of the equipment needed.
2. Construction of the apparatus.
3. Final connecting, assembling, wiring, etc.
4. Preliminary testing and alignment.
5. Test runs.
6. Modification and supplementation as necessary.
7. Handling and interpretation of data.

These operations will be discussed in the order listed.

Sketching and Drawing

The introduction of photography was once believed to have eliminated the need for drawing and painting. The high speed and accurate depiction made possible by the camera would seem to have made manual graphics obsolete. It is now obvious that this has not been the case. So far, no camera has been devised which can photograph a concept or record a memory. The creative individual can essentially "build a model" on paper and add or subtract from it as he deliberates.

He can then "move about" this model, looking at it graphically from all sides, in cross section, or inside out, as he pleases. If he once saw a device but it no longer exists or is no longer accessible to him, he can still reconstruct its shape and form on paper. A drawing can depict a static situation or full motion. It has no limits as to scale or scope. Many successful inventors owe part of their good fortune to a high degree of artistic ability, which enabled them not only to rapidly visualize a new concept but later to present it effectively to potential backers, licensees, or purchasers.

The inventor who is not already trained to some degree in the principles of drafting should be aware of the three most useful forms of graphic depiction: the schematic, the perspective, and the working representation. The schematic drawing makes use of standardized symbols to show equipment, processes, chemical compositions, mathematical relationships, etc. The perspective drawing is a representation of what the device would look like if it were actually built and photographed. It is realistic art and often employs techniques such as shadowing to increase the illusion of three dimensional representation. The working drawing is a planar representation of the object from one, two, or more sides and indicates details of construction. The working drawing can be full sized or to scale, enabling an artisan in either case to construct the device from direct measurements. A working drawing will often contain notes to the constructor advising on materials, tolerances to be observed, surface finish, tests to be made on components, and miscellaneous data of value or interest to its users.

An example of the three types of drawing applied to an inventive concept is shown in Fig. 7-1. The inventor has in mind a voltage divider which can be used for indicating angular position. It has very low frictional resistance because it uses a magnetically-held ball instead of the usual slider. Fig. 7-1A, a schematic, indicates how the device operates. A wound resistance is contacted by a spherical armature which also rubs against a "bus" wire. The ball is caused to change position by moving a magnet near it. A fixed voltage is applied across points b and c and an output device, such as a voltmeter, is connected between points a and b. With the ball at point c, the

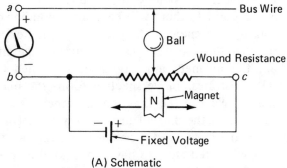

(A) Schematic

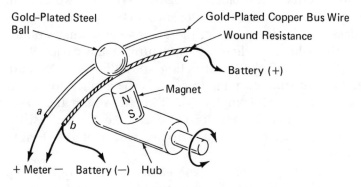

(B) Perspective

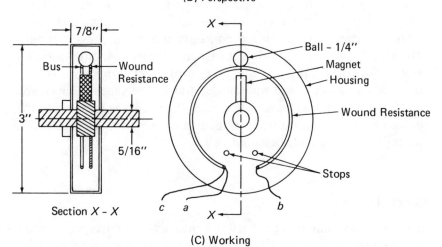

(C) Working

Fig. 7-1. Types of drawings.

meter will read the full value of the applied voltage. As the ball is moved toward point *b*, there will be a corresponding decrease in the voltage read by the meter. At point *b* the meter will read zero. The reading can thus be used as a measure of the angular position of the ball, the magnet, and any device attached to the magnet. Figure 7-1B illustrates, in perspective form, the placement of the essential parts. We can see that the movement is circular, the ball is attracted towards the magnet in such a way that it exerts contact pressure on both the wound resistance and the bus wire. The ball is made of gold plated steel to have both good magnetic properties and low contact resistance. Figure 7-1C is a working drawing. The essential dimensions are indicated. The right hand view looks at the device from its open end; the other view is what would be seen by cutting along the line *X-X* and turning the device 90°. Additional features of the construction are indicated. A housing prevents the ball from escaping (a cover to be applied to the open end will seal off the ball completely). Two stops limit the travel of the magnet to within the angular range covered by the resistance. A craftsman could construct a working model of the device by the use of Fig. 7-1C.

It is, of course, possible to combine two types of drawing. A perspective type of working drawing can be made showing dimensions, materials, etc. To indicate hidden parts, the perspective sometimes depicts a device with certain parts removed — the so-called "exploded" view.

The drawing up of invention concepts and the progressive modification of these drawings in accordance with subsequently-acquired information, calculations, etc., is a form of experimentation. By virtue of its low cost and wide range, drawing becomes an invaluable tool for use by the inventor. It would, therefore, be most profitable for those who are not already skilled in drawing techniques to take courses in drafting and sketching.

Model Making

Invention modeling can be classified into six general types: experimental, evolutionary, skeleton, working (or prototype), presentation and production.

The experimental model may bear no resemblance at all to the final embodiment. An improvement in a water softening method may be developed experimentally in glass vessels and tubing. The process, when finally perfected, is carried out in open channels of flowing water.

An evolutionary model is designed to permit rapid additions, subtractions, and modifications as the invention is improved. An electronics breadboard is one type of evolutionary model. If properly designed, components can be easily added or removed (in some cases without the use of solder) and two or more breadboards can be readily coupled together.

As might be expected from its name, a skeleton model incorporates only the barest essentials of the device being developed. Other components are eliminated to permit easy access and observation of the experiment. An example would be an automobile frame containing only seats, motor and steering functions for use in evaluating the effectiveness of various seat belt constructions.

A working model may or may not resemble the final device. It does, however, successfully demonstrate the principle of the invention. A new type of writing pen contains a system for the electrolytic decomposition of a transparent, metallic dye solution. When the latter is decomposed, it turns deep blue. One of the electrodes is also the writing point, so that the decomposed dye can be applied to paper. The inventor builds a working model, comprised of a flashlight battery, a toggle switch, an eye dropper, two electrodes, and the associated wiring. One of the electrodes projects a short distance through the end of the eye dropper. The assembly demonstrates the principle and can be used to produce samples of writing. The final version will require miniaturization, the selection of a tiny battery having sufficient capacity, the design of a special switching device which remains on only when the user exerts pressure during the act of writing, and a suitable cartridge containing electrolyte.

The presentation model is a selling tool. It not only demonstrates the principle of the invention but also gives an impression of the final product. The degree of sanding, polishing and painting is sufficient to produce a pleasing impression. An example would be a new type of speedboat in which the passenger-carrying body is joined by a spring suspension system to two flotation pontoons. The latter also support the motor and propeller assemblies. Pitching, rolling, and wave-induced motions are not transmitted to the body. The inventor builds and tests a working model. He then has a finely detailed scale model constructed. This is his presentation model, to be used in raising capital for commercial production of the invention.

A production model demonstrates the alterations which must be made in previous versions to permit economical manufacturing of the new device. Whereas the working model or the prototype may have used expensive and custom made parts, the production model is built with competitively priced components. The production model utilizes mass produced items and can also be used to set up preliminary quality control procedures. A new kind of typewriter, for example, might be initially modeled with cast gears and a steel frame. The production model is made of stamped gears and a molded plastic frame.

In the construction of any of the above types of models, many fabrication processes are available. It will be obvious that certain of these are of lesser value in a particular case than others, inasmuch as they may require elaborate and expensive equipment which is usually only justifiable in large scale manufacturing.

For the manipulation of solids such as bars, rods, planks, sheets, and wires, the following methods are used:

1. The cutting processes: sawing, turning, grinding, milling, stamping, chemical etching, spark machining.
2. The forming processes: bending, rolling, swaging, extrusion.
3. The melting processes: casting, sintering.
4. The joining processes: nailing, riveting, gluing, soldering, welding, bolting, threading.
5. The heat treating processes: hardening, annealing.
6. The surface treatment processes: painting, electroplating, anodizing, chemical preparation.

For the handling, makeup, or conversion of soft materials such as fabrics, fibers, leather, rubber, etc., the following are generally employed:

1. The interlacing processes: weaving, braiding.
2. The cutting processes: shearing, stamping.

3. The joining processes: sewing, riveting, gluing, fusing.
4. The forming processes: bag molding.
5. The lay up processes: incorporating with adhesives and mold-
    ing, papier-mâché techniques.
6. The fiber orientation processes: stretching, pressing, shrinking.

In choosing one or more of the above for the preparation of a particular model, a number of general principles will be of help. It is desirable to fabricate the model of standard parts whenever possible. Use standard rods, beams, wheels, gears, springs, etc., to cut down fabricating time and expense. Special hardware available for replacement purposes is often an excellent source of parts for an invention. Radio hardware or sliding door parts, for example, will provide knobs, pulleys, gears, and levers which can usually be adapted to the particular construction at hand. Automobile and refrigerator parts stores can supply components for the construction of hydraulic systems — tubing, pumps, valves, etc. Military surplus is usually worth disassembling to build up a store of valuable components. Manufacturers of optical equipment must work to exacting standards. Lenses, mirrors, and prisms, rejected because of minor imperfections not directly related to function, are often available in a wide range of sizes. The catalogues of very large hardware stores list many kinds of standard parts.

A second general principle in the construction of models is to build the unit up from individual pieces, rather than cut it to shape from a solid block. In production, the machining away of a large amount of metal is wasteful, but at least the chips are of sufficient quantity to make them valuable as scrap. In one-of-a-kind construction, the removal of large volumes of material is seldom justified. The design can usually be altered to permit the device to be bolted or welded together from standard pieces.

A third principle in the building of models is to use the most easily fabricated material which is consistent with the desired result. A cardboard model may demonstrate the feasibility of an invention in an elegant manner and even permit the taking of useful data. To employ any more substantial material in this case would represent wasted construction time. A coil of 3/32 in. steel wire (the diameter used in coat hangers) and a simple bending jig are good investments for the inventor. Many kinds of apparatus can be fabricated from the wire once the experimenter starts to think in terms of this very versatile material. Supports and frames can be formed from a single piece of wire. Holders, clamps, wheels, and even gears can also be fabricated of wire.

A fourth principle in model building is to employ optimization techniques. These are useful where it is desired to obtain precision without the need for great mechanical skill. Kinematic design, one method of optimization, is based on minimizing the area of contact between surfaces or minimizing the number of restraints and thus reducing the amount of skill necessary to generate accurate contact surfaces. A four legged stool may rock if the workmanship is not of the highest quality; a three legged stool is always stable. If the legs of the stool terminate in flat surfaces, they may not make uniform contact with the floor. This places uneven stress on the various parts of the stool when it is under load. If the legs terminate in spheres, on the other hand, contact with the floor is reduced to a very small area and is the same for each leg. If it is now desired to place the three legged stool (or a platform holding an instrument) in an exactly reproducible position, it is only necessary to cut three, non-parallel, V-shaped grooves into the floor in a position such that each accommodates a spherical leg end (Fig. 7-2A). The stool can be removed and then replaced into the same position with a high degree of accuracy. The spherical ends can be made of steel bearing balls; only average precision is required in cutting the grooves. If the grooves are made parallel, the stool can now be moved along in one direction with almost no play or wobble. A somewhat simpler arrangement is to cut the grooves into a block and utilize cylindrical bars, as shown in Fig. 7-2B. This construction would be a good substitute for the more difficult method of scraping and honing large flat surfaces as is done in some kinds of lathe bed design.

Many mechanisms involve the limited motion of one part around another. Normal hinges have some play even when new. When the motion required is relatively small, hinge play can contribute a large error. "Spring hinges," as shown in Fig. 7-2C, made of flat strips of resilient metal (phosphor bronze, e.g.) have almost no play and serve very well for this type of application.

Some optimization techniques are derived from a realistic appreciation of operating problems. In the case of a wheel or gear which is

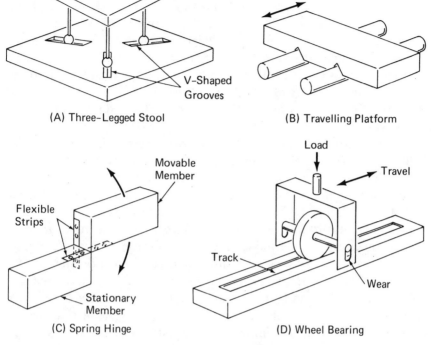

(A) Three-Legged Stool

(B) Travelling Platform

(C) Spring Hinge

(D) Wheel Bearing

Fig. 7-2. Optimization of design.

carrying a load over a track (Fig. 7-2D), the designer has the choice of allowing the wheel to turn free on a fixed axle or fixing the wheel and allowing the axle to turn in end bearings. If the wheel turns free, wear will cause enlargement of its center. This will result in loss of precision because the wheel must now turn around a changing center. It will also be prone to wobble. If the second option is followed, the bearing supports become elongated but no loss of precision occurs.

The above are only a few examples of optimization techniques. For a more thorough discussion of this valuable method, the reader should refer to the texts listed at the end of this chapter.

A fifth principle in model making is concerned with making provisions for error minimization. It is desirable that the means for making a measurement be frequently checked; in this way, the possibility of acquiring large amounts of invalid data when testing the model is reduced as much as possible. If, for example, the model's performance is critically dependent on the amount of a certain chemical present in a solution, the analyzer which measures concentration must be periodically checked. One way of doing this is to include in the model a test tank containing a weighed amount of the chemical dissolved in very pure solvent. At regular intervals, the flow of solution from the process to the analyzer is interrupted and replaced by a flow from the test tank. If the analyzer does not read correctly, it is adjusted before it is returned to service. The frequency of adjustment will depend on how far the reading differs from the correct one and the time elapsed from the previous check. In some cases, it will be necessary to check the analyzer with several test tanks, each containing a different concentration of the chemical.

Another method for error minimization useful with some models is that of shielding. Certain disturbances may cause erratic operation. Vibration transmitted from the surroundings (or from its own drive system) may cause undesirable movement of critical parts. Heat may change the characteristics or ratings of certain components. Stray electrical signals from power lines may enter sensitive measuring circuits. The model can be isolated from most disturbances once these have been identified. Elastic mounting will greatly reduce the transmission of vibration; insulating layers or large area radiators will reduce the effects of heat. Woven metal cloth which is placed around electrical apparatus and then grounded will effectively shield the model from stray electrical fields.

A sixth and final principle is to seek qualified help whenever a seemingly insurmountable obstacle is encountered. The task of model fabricating is unpleasant for some inventors even though they enjoy the testing phase. After one or two unsuccessful experiments with improperly designed models, these inventors will find it easy to put off further work. The construction of an adequate model by others removes this source of discouragement.

Measurement Taking

Measurements are the lifeblood of the invention process. Much of the technical proliferation of today as compared to the modest progress of previous times lies not in the fertility or quality of the imagination possessed by modern inventors but in the increased sophistication of measuring devices. Once a phenomenon has been measured and the measurement repeated, it can be verified by others. The data can be worked over by certain mathematical techniques to extract additional information. The process as carried out in the experiment can be scaled up or down, as desired. Other mathematical methods permit educated guesses to be made as to the best values of certain variables to give optimum performance of the device. Finally, the results of the measurements can be used to design an end product which will bring the benefits of the invention to large numbers of people.

Measurements can be characterized by the types of energy with which each is associated. Thus we have instruments and techniques for gauging the flow of mechanical, electrical, chemical, nuclear, optical, acoustic, and thermal energy.

A survey of instrumentation shows many conversion methods — i.e., the particular form of energy to be measured or particular phenomenon to be detected is often "transduced" to another, more easily handled form. In Table 7-1 are listed the kinds of measurements connected with each form of energy. An additional measurement, time, is often associated with many of the others and enters into the computation of parameters such as power, speed, frequency, and various "transfer" constants.

TABLE 7-1. TYPES OF MEASUREMENTS.

Energy Form	Measurements Commonly Made
Mechanical	Position, dimension, weight, volume, pressure, speed.
Electrical	Voltage, current, resistance, power, capacity, inductance, phase, spectral content.
Chemical	Equivalents of one substance for another (acid for base, etc.), chemical potential, composition.
Nuclear	Particle energy, identity, velocity, capturability, momentum.
Optical	Color, absorption, transmission, wavelength, phase, spectral content.
Acoustic	Power, absorption, transmission, waveshape, spectral content, phase.
Heat	Temperature, radiation, conduction, convection, thermal capacity, heat transfer properties.
Time	Intervals, frequency phenomena.

Mechanical measurements represent the most basic single class. Length, for example, is formally defined as a multiple or a fraction of the distance between two scratches on a certain platinum bar with the latter being held at a specified temperature. The primary, "standard" bar is kept in a laboratory in Paris.* The separation between the scratches has been copied (and subdivided for convenience) on millions of metal, plastic, wooden, and glass strips. When we say that an object is 32 meters long, we mean that if a divider is set between the scratches on the bar and used to step off the length, it will require exactly 32 steps. The directness and easily grasped significance of this concept has resulted in its almost universal application to most other measurements.

Whenever possible, we convert temperature, voltage, power, etc., to a proportional mechanical motion. The position of a pointer from some starting point (i.e., the partial length of a scale) is then a measure of the quantity we wish to investigate. If a pointer swings completely across the scale of a voltmeter for an input of 200 volts, it is desirable that it move halfway across when 100 volts is applied, and one quarter of the way with 50 volts, etc. If this is not the case, we can still make use of the meter by varying the input voltage by known steps and marking in the corresponding positions of the pointer. For values in between any two calibration marks, linearity is assumed and we accept the fact that there will be some error as a result. In Fig. 7-3 are shown two scales. The upper one indicates a linear relationship between input (voltage) and output (pointer position). The smallest subdivisions are equally spaced at all points along the scale. In the scale shown in the lower illustration, the divisions become more crowded towards the left end. To construct the upper scale, it was only necessary to apply accurately known voltages at a few points; e.g., 50, 100, 150, and 200, and then subdivide the scale evenly. To construct the lower scale, it was necessary to apply exact voltages for each subdivision. In taking the reading indicated by the pointer in Fig. 7-3 B, it is assumed that the distances between

---

* The original separation between scratches has been measured in terms of the number of wavelengths of a monochromatic light source. The standard meter is, therefore, an "absolute" standard because it has been defined in terms of a basic property of nature – the speed of light.

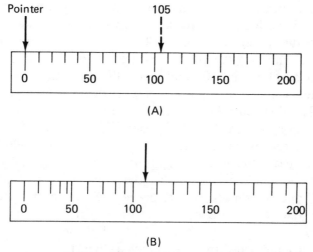

Fig. 7-3. Meter scale fundamentals.

calibration points is approximately linear; the reading is therefore estimated at 105 volts. This number will be in error by the amount of deviation from linearity.

In some instruments, the scale is non-linear at either end but linear at the center. It is generally recommended by the manufacturer that the conditions of the measurement be adjusted, if possible, so that only the center section of the scale be used with these instruments.

Measurements and the devices employed for their realization are characterized by several important properties. One of these, linearity, has just been discussed. The others are: sensitivity, dead zone, speed of response, and noise.

Sensitivity can be defined as the amount of pointer movement obtained per unit of applied energy. If the usable part of the scale of Fig. 7-3A is 10 in. wide, the sensitivity of the instrument is 10 in./200 volts or .05 in./volt. It is obvious that this meter would be very hard to read for voltages less than .5 volt or for similarly sized changes in higher voltages — e.g., a change from 150 to 150.5 volts. If it becomes necessary to read .05 volts, a meter with a full scale range of 1 volt would be a logical substitute. The sensitivity would now be 10 in./ volt; a .05 volt signal would then produce an easily read 0.5 in. deflection. If the problem is to detect a .05 volt change in a large voltage, a differential method (to be described shortly) is used.

The effect of dead zone comes into play if we approach a particular reading from both ends of the scale. If we add 1 volt sources one at a time to the meter input until 100 volts is applied, the meter will stop a little short of the 100 indication. If, on the other hand, we start with a voltage of 200 and gradually decrease it to 100 volts, the pointer will stop at some value higher than the true reading (Fig. 7-4A). The difference between the gradually-approached "up" and the gradually-approached "down" reading is the dead zone, a measuring error caused by bearing friction, uneven magnetic fields, etc. In a practical case, the sudden application of a voltage may cause a reading error on either side of the true value. If, for example, 100 volts are applied to most meters, the pointer will overshoot the 100 mark and then oscillate one or more times. In Fig. 7-4B is a plot of the variation in reading versus time. If the last oscillation is such that the pointer approaches from the up side, then the final reading will

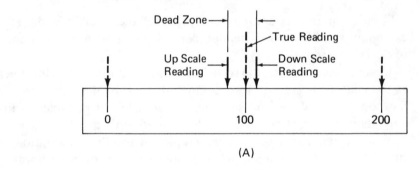

(A)

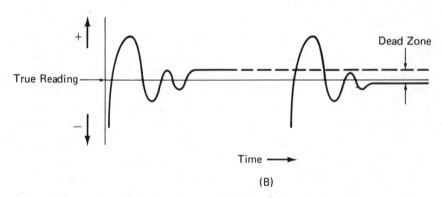

(B)

Fig. 7-4. The "dead" zone.

be high. If the last oscillation approaches from the down side, the reverse will be true. In many instruments, dead zone has been reduced to a negligible quantity by good design and the use of special materials for bearings, magnets, and springs. In certain developmental systems, on the other hand, the inventor will find dead zone to be an important factor. The most effective method for decreasing the effect of dead zone in these cases is to take many readings and average them.

Speed of response of a measuring device can be defined as the time that the device requires to respond to a suddenly applied signal. Speed of response is of little concern when we have a slowly varying signal. The pointer overshoots, oscillates, and finally settles down to some value which is within the dead zone of the true reading. If, however, the signal changes rapidly and continuously, it can be seen that a condition could be reached where the pointer never attains its equilibrium value for any of the signals and all readings become erroneous. Speed of response in the case of a pointer type of voltmeter is related to such factors as the mass of the moving system, the strength of the return spring, and the rapidity with which the magnetic field in the instrument's coil can build up and collapse. With rapidly changing signals, a human factor is also introduced. Even if the instrument can follow these changes, it becomes increasingly difficult for the experimenter to write down the values as fast as they are presented on the scale. Certain means can be employed to overcome slow speeds of response and human limitations in recording. Meter movements designed for higher speeds make use of miniature components and lightweight materials. The pointer is made very short, as is scale distance, so that full scale travel time is reduced. The pointer may be provided with a tiny pen and made to press against a moving ribbon of paper to automatically record the change in signal. At still higher speeds, a cathode ray oscillograph is used. The nearly weightless electron beam is employed as a pointer and permits the accurate depiction of signals varying at extremely high speeds.

Noise places an ultimate limit on the sensitivity available in any measurement. Noise becomes important in those types of development where the experimenter must measure extremely small quantities. If, for example, he is concerned with a new type of heart pacemaker, he is dealing with minute voltages. To read these on a meter, he must first amplify a microvolt signal to the volt level. Random electron motion in the wires and transistors of his amplifier, atmospheric electricity, pickup from power lines, and even cosmic rays produce voltages almost as large as those he is attempting to amplify. The result is a large signal of which only a small percentage is varying in accordance with the biological effects he is studying. Noise effects come into play with any other type of amplification as well. As an illustrative example we will choose an experiment which involves the measurement of very small pressures. To amplify these, we will use the "optical lever" scheme shown in Fig. 7-5. A very thin rubber sheet is used to seal off a cup-like structure. A tube connects the pressure-producing process to the bottom of the cup. Slight pressure changes cause minute deflections of the rubber sheet. A very light "T" bar rests with one leg on the rubber sheet and with the second leg on the non-moving lip of the cup. A mirror is attached as shown and reflects a thin pencil of light onto a screen placed at some distance X_1 from the cup. Small deflections

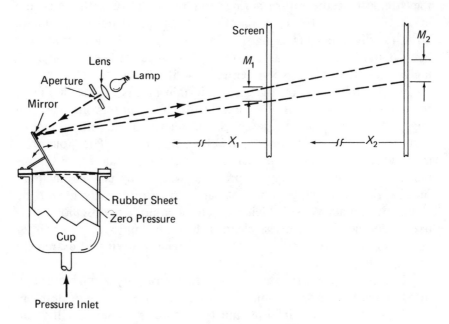

Fig. 7-5. Amplification and noise.

of the sheet will produce tiny movements of the mirror. At the distance X_1, these movements will produce amplified deflections of the spot of light appearing on the screen. Movement M_1 can be many times greater than the minute deflection of the rubber sheet; the arrangement can, therefore, be used to measure very small pressure changes. If the screen is moved further back to the distance X_2, the movement M_2 becomes even greater. As the distance is increased, however, it is observed that the length of the illuminated spot appearing on the screen also increases (Fig. 7-6). The greater length is the result of vibration of the diaphragm (even though pressure is not changing). The vibration is caused by the random impact of air molecules on the surface of the rubber sheet, drafts, earth-borne disturbances, etc. As the optical amplification is increased, the apparatus becomes sensitive enough to detect more of the noise which surrounds it. Since the increased length of the spot of light decreases the certainty of the readings, no real benefit is obtained (after some critical point) by increased amplification. Noise has again limited the ultimate sensitivity of the measurement.

It is possible, in some cases, to devise filters which will partially remove noise and thus permit some additional amplification. Other methods for reducing the effects of noise include shielding (for electrical signals) and the lowering of temperature. Shielding techniques, as was mentioned earlier, employ a metallic fabric which is

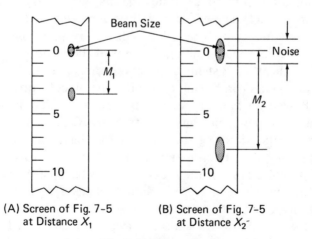

(A) Screen of Fig. 7-5 at Distance X_1

(B) Screen of Fig. 7-5 at Distance X_2

Fig. 7-6. Effect of noise.

drawn over signal-carrying components of the measuring system and then grounded. Extraneous noise is thus safely bypassed to ground before it is able to mix with the useful signal. Very low temperatures decrease random molecular motion. Where this method can be used, it is very effective in decreasing the effects of noise.

Transduction

In the above discussion, frequent reference was made to voltmeters in order to illustrate basic principles in the reading and interpretation of scale indications. In actual practice, it is found that voltmeters can indeed be used as the final element in almost any measurement. The highly advanced development of transducers — devices for converting many different kinds of inputs to proportional electrical voltages — makes the voltmeter the most versatile of output devices. In the measurement of temperature, for example, a thermocouple is often used. This is comprised of two wires of differing metals welded together at one end to form a bead. The latter, when heated, generates a small voltage which is proportional to its temperature. The basic circuit is shown in Fig. 7-7A.

In another type of transducer, position is converted to an electrical signal by the method shown in Fig. 7-7B. The object under study is rigidly coupled to a movable iron core of a specially designed transformer. Since the core concentrates the magnetic energy produced by the primary coil, changes in core position will greatly alter the amount of secondary voltage obtained. It is therefore possible to calibrate the voltmeter reading in terms of the position of the object. This indicating system is quite versatile, inasmuch as many other measurements can be made with it. The object whose position is being followed on the voltmeter can, for example, be the pan of a spring balance, a newly-machined surface, the diaphragm of a pressure indicator, the vibrating surface of an airplane wing under test, the movable jaw of a micrometer, or the piston of a hypodermic syringe. The voltmeter, in these cases, would become, respectively, a readout for a weighing machine, a flatness checker, a pressure gauge, a vibration meter, a length measure, or a volume indicator.

Many other kinds of transducers have been developed so that most measurements can be carried out with a transducer-voltmeter

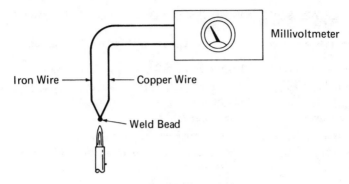

(A) Simple Thermocouple

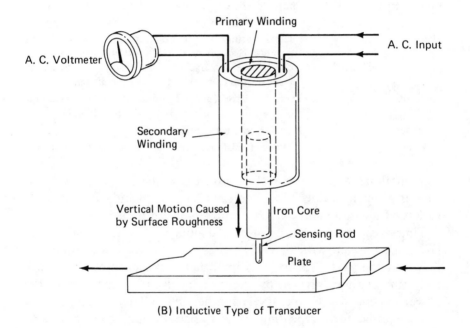

(B) Inductive Type of Transducer

Fig. 7-7. Transducers.

TABLE 7-2. SOME COMMONLY USED TRANSDUCTION METHODS.

Property to be Measured	Transducer Used To Drive Voltmeter
Light transmission	Photoelectric cell
Nuclear radiation	Geiger counter, proportional counter, photoelectric scintillator
Acoustic power	Condenser microphone
Liquid level	Float + movable core transformer
Radio signal power	Thermocouple (radio frequency voltage is first converted to heat)
Rotary speed	Electrical generator
Magnetic field	Rotating coil
Color intensity	Photoelectric cell + filter (to remove all colors except the one being studied)
Odor	Special battery (air containing the odor-producing substance is bubbled through a liquid near one electrode of the battery. Voltage produced depends on concentration of the substance in the air sample)
Humidity	Special resistor (two electrodes imbedded in a ceramic which absorbs moisture from its surroundings)
Thickness	Movable core transformer
Gas flow	Hot wire (resistance varies as it is cooled by flow)

combination. A partial list of transducers which can be used in this way is given in Table 7-2. The samples chosen indicate the wide variety of measurements which are possible with transducers.

Time is a parameter often measured along with other variables and later incorporated with them. There is a large class of measurements in which the "rate" concept is important. For example, the work put into a system divided by the time involved is a measure of the power expended. Frequency is another time related concept. The time required for a repeating effect to pass through one cycle is an important number in characterizing many phenomena. Time is measured as a whole or fractional multiple of the period required for the earth to turn through an averaged day. All methods for time determination make use of vibrating bodies whose motions are accurately counted. A tuning fork, oscillating balance wheel, quartz crystal, or even the oscillating atoms produced by electrical discharge in gases have one characteristic in common: their vibrational periods are very reproducible. It is found that some of these periods are even more accurate for gauging time than the period of the earth because the latter is slowing up with each day.

When measuring reasonably long periods of time, the inventor has a choice of straightforward methods — watches, electrical clocks, and radio signals. When timing very short periods, however, he must employ more sophisticated gear: the triggered oscilloscope, the high speed photograph, the optically switched timer, and the stroboscope, among others.

Measurement System Design

The exercise of a number of precautions in the design of a measuring system will often spell the difference between observing an effect (and being able to develop the corresponding invention) or obtaining a non-conclusive result. Modest and low cost instruments will do an elegant job if properly applied; expensive and elaborate devices can fail miserably if the user is not aware of certain instrument limitations. We shall discuss three of the most important concepts in measurement design.

Perhaps the most general rule of all measurement is that of "impedance matching." The term is borrowed from electrical theory but applies to all measuring techniques. An instrument must, for minimum error, be matched to the system under study so that the act of measurement does not withdraw (or add) enough energy to distort the system and thus create an error. Suppose we wish to measure the length of a rubber block with calipers. As the jaws of the calipers touch the block, compression takes place — the process of measurement is actually changing the variable being measured. The "impedance" of the rubber block — i.e., its resistance to distortion — is low in comparison with that of the jaws of the calipers, so we have a mismatch. A "softer calipers" would be a light beam which would cast a shadow of the block against a calibrated screen. In making electrical measurements, experimenters will sometimes use a voltmeter which draws a considerable amount of current in comparison with the current available in the circuit being measured. The result is an erroneously low reading. If an amplifier is interposed between the test points and the voltmeter, the signal strength can be raised enough to permit operation of the meter without undue loss. If liquid flow is being measured by means of an in-line turbine and if the latter is improperly designed, a significant percentage of flow energy is being used to move the turbine blades. The result is a decrease in flow and inaccurate measurements. Impedance matching, in this case, would consist of decreasing the weight of the turbine blades and improving the quality of the instrument bearings. Another example of impedance mismatch may occur in the measurement of vibration. If the instrument is not tuned to be resonant at the frequency of the vibrating system, some vibrations will be reflected back into the system at the point of contact with the instrument and readings will be erroneously low.

A second basic consideration in measurement is that of compensation. When we state the density, electrical resistance, length, or color of an object, we make certain stipulations about temperature, atmospheric conditions, local gravity, etc. If any of the latter factors is different at the point where the number is to be used than it was when the original measurement was made, some method for compensation must be employed. In simple systems, compensation can be done by calculation. If a certain steel rod is found to be 12.25 in. long at 82° F, its length at 32° can be calculated and comparison made with the measurements of others (which were also compensated to 32°, the freezing point of water). In many cases, however, there are too many variables to permit the convenient use of computation. A most effective compensation method is to employ a standard with which to compare the unknown. If, in the above example, we had available a second steel rod which was exactly 12 in. long at 32° F, we could compare our unknown with it at room temperature. Expansion effects would be the same for both rods and no calculations would be required. Several specific examples of comparison methods are illustrated in Fig. 7-8. In making precision weighings, a number of factors normally overlooked must be considered. If the object to be weighed is of low density so that it occupies a relatively large volume, the air it displaces has a buoyancy effect which will result in a low reading. In the comparison method shown in Fig. 7-8A, the object is first counterbalanced by a lesser but known weight of similar density. Air buoyancy effects on the weight and on the object act on opposing pans of the balance and thus cancel one another. The final weight is determined by a slider, which is moved until balance is

reached. The position of the slider then indicates the difference between the weight of the object and the standard. In Fig. 7-8B is shown
a comparison method for voltages. A battery or power supply whose
voltage is accurately known is connected across a voltage divider. By

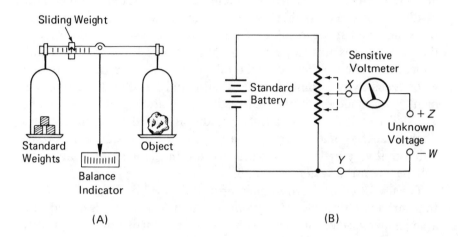

(A)   (B)

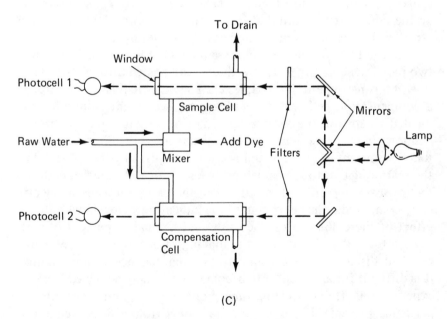

(C)

Fig. 7-8. Compensation methods.

moving the slider, any voltage between zero and that of the battery is obtainable. The slider can be accurately marked off in terms of the voltage appearing between points X and Y. The unknown voltage is applied between Z and W, with its polarity opposing that of the adjustable source. A sensitive voltmeter acts as a "referee." When the divider is adjusted so that the voltmeter reads zero, the position of the slider indicates the value of the applied voltage. The linearity of the voltmeter is not important because it acts here only as a null indicator. The comparison method permits the use of a voltmeter whose sensitivity is much greater than could be used if the measurement were made directly. As a result, the matching potential can be adjusted to within the smallest scale division of the voltmeter. As a final advantage of the comparison system in this instance, it is to be noted that very little power is drawn from the source of the unknown voltage when the system is at null.

The use of compensation in an optical device is shown in Fig. 7-8C. In a certain pollution-control monitor, water from a river is mixed with a small percentage of dye. The resulting color is photoelectrically gauged. The more acid which has leaked into the river, the deeper will be the color. It is found that fine, suspended silt in the water causes errors in the measurement because of light scattering. In the compensation method shown, the river water sample is continuously divided into two parts. One half is treated with dye and passed through the "test" side of the instrument, the sample cell. The other half is passed through a compensation cell without addition of dye. Light from a lamp is divided by mirrors, passed through color filters and directed through windows in both the sample and compensation cells. Photocell 1 registers the amount of light remaining after absorption by the color and by the scattering due to suspended particles. Photocell 2 registers the effect of scattering only. The outputs of the two photocells are electrically compared. The net output of the indicating meter shows only the effect of the color and can be calibrated to read in percent of acid.

As a final consideration in the making of accurate measurements, we shall discuss the establishment and maintenance of impartiality. It is difficult for an inventor to become emotionally detached from his experiments. He is investing time, hope, money, and work into his particular project. His mental state may range from high excitement to one of intense urgency. Under pressure he may take unjustifiable shortcuts, draw incorrect conclusions, or transpose data. Cases of the latter have occurred even when the experimenters were highly trained. Some years ago, an important chemical project was being carried out jointly by a number of corporate laboratories. To maintain analytical accuracy, check samples were periodically circulated among the cooperating companies. The test consisted of submitting to each lab ten known samples for acid titration (a process in which a chemist slowly adds acid to a measured quantity of the sample material until the latter changes color). On one occasion, nine samples were made identical but one was diluted to have an acid requirement 1 cc less than the rest. The "erroneous" sample was placed in the seventh position in the sample holder. A properly analyzed set might have been reported as follows: 9.65 cc, 9.63, 9.71, 9.70, 9.64, 9.67, *8.69,* 9.63, 9.73 and 9.65. In the actual test, every lab reported *all* the first figures as 9. Seeing the same first number for six successive tests subconsciously convinced the participants that all were the same.

Projects which involve human subjects are doubly sensitive to the effects of bias and emotional reaction. Thus, the suggestibility of the subject may combine with the anxiety of the experimenter to produce a completely erroneous result. If, for example, the inventor is using a group of his relatives to test a fuel saving device for automobiles, he will need to examine carefully all phases of the experiment, including his own desire to succeed and his relatives' wish to please him.

A number of methods are available for minimizing the effects of bias in making measurements. The most obvious of these is repetition. In many tests, sources of bias will be discovered in the mere act of duplicating the measurement. A second method is to let the measurement be made by others. Habits which may be causing a systematic error will not be shared by different experimenters. A third method is randomization. If the experimenter wants to find the effect of added weight on the acceleration of a certain vehicle, for instance, he should not proceed in one direction — i.e., increase the load and measure the result. Even though a one-directional experiment is the most convenient and will give a smooth graph of acceleration vs. weight, it may induce error. It will not, for example, reveal the effects of semi-permanent bending of the vehicle frame with increased load. If he studies acceleration with a large load, then a small one, and continues with randomly selected loads, the curve obtained will not correlate smoothly and will show up the shortcomings of the test vehicle. The experimenter will then be led to the construction of a more suitable testing device. In the acid titration test described above, randomization of the samples would eventually have brought the 8.69 cc sample into the number one position, where its true value would have been recorded.

Data Handling

Measurements produce numbers; it is from these that the inventor draws conclusions, plans further work, designs practical equipment, or gathers arguments to convince others that he has made a worthwhile discovery. The numbers generated from an experiment can be put to these uses in a number of specific ways. The inventor may employ his data for any of the following:

1. To see if a certain effect really exists.
2. To determine the repeatability of the effect and to establish a "confidence" level in the system and in the measuring tools.
3. To predict what might happen if some experimental condition (amount, size, speed, temperature, etc.) were to be extended beyond previous limits.
4. To predict what would have happened in the intervals between the actual experimental conditions employed. One possibility here would be to determine if an area of optimum results was overlooked because of the particular conditions chosen.

To study these uses, let us stipulate the following investigation: an inventor wishes to explore the structural possibilities of a material composed of aluminum fibers imbedded in epoxy resin. He reasons that a composite of this kind would have a number of advantages over fiberglass-epoxy, including increased ductility and better heat conductive properties. He is interested first in finding the tensile strength of various compositions of aluminum in epoxy to see whether such a material would have a reasonable chance in competing with the older composite. He starts with a 5% by weight

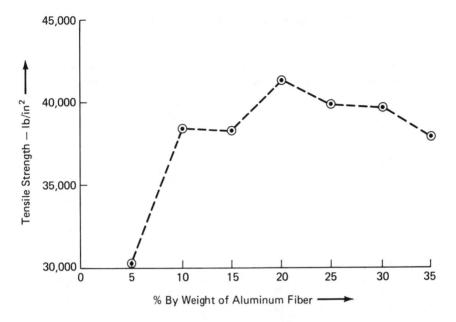

Fig. 7-9. Strength of certain epoxy-aluminum fiber compositions.

of aluminum and obtains a tensile strength of 30,200 lb/in.^{2}. This is considerably better than the epoxy alone. He now begins to experiment with other percentages. A graph of his results — tensile strength vs. % by weight of aluminum fibers — is shown in Fig. 7-9. It appears that some improvement occurs with increasing fiber content but it is difficult to draw definite conclusions. The inventor now chooses one of the "better" tests — that found at 20% aluminum — and carefully runs 15 repeats. His results are shown in Table 7-3. The average of the 15 tests is 39,500 lb/in.^{2} (593.2/15 x 1000). In the third column are listed the deviations of each test from the average, along with a plus or minus sign to indicate the direction of the deviation.

The average deviation (from the average) is found to be ±1.7. The latter figure indicates that any one value in the table is uncertain by this amount. Test 7, for example, may have turned out to be any value between 38,400 and 41,800 lb/in.^{2} and remained within the average accuracy of the test. The experimental procedure by which the fibers are imbedded in the resin, the apparatus with which the

TABLE 7-3. TENSILE STRENGTH OF ALUMINUM FIBERS IN EPOXY — 20% BY WEIGHT.

Test Number	Tensile Strength (thousands of lb/in.^{2})	Deviation From Average	Square of Deviation from Average
1	42.1	+2.6	6.8
2	37.2	−2.3	5.3
3	40.2	+0.7	0.5
4	38.6	−0.9	0.8
5	39.7	+0.2	0.0
6	39.1	−0.4	0.2
7	40.1	+0.6	0.4
8	43.3	+3.8	14.4
9	36.1	−3.4	11.6
10	38.2	−1.3	1.7
11	40.1	+0.6	0.4
12	41.4	+1.9	3.6
13	36.7	−2.8	7.8
14	38.1	−1.4	2.0
15	42.3	+2.8	7.8
sums	593.2	±25.7	63.3

Sums ÷ 15: 593.2/15 = 39.5 = average

25.7/15 = ±1.7 = average deviation from average

63.3/15 = 4.2; $\sqrt{4.2}$ = ±2.1 = standard deviation

test samples are pulled apart, or, perhaps, the variable shrinkage in the sample as the plastic cures (a decrease in the effective area) are factors which prevent the overall reproducibility from being better than ±1700 lb/in.^{2}. A more accurate approach to the calculation of the average deviation involves squaring the individual deviations, adding them up, dividing by the number of tests, and extracting the square root of the quotient. The result is the so-called "standard" deviation, which has certain characteristics useful in statistical calculations. Details of these can be found in the texts listed at the end of this chapter. In the present case, the standard deviation is approximately ±2100 lb/in.^{2}.

The inventor next repeats each test shown in Fig. 7-9 several times (e.g., five repetitions). He randomizes the test order: 15, 25, 5, 30% aluminum fiber, etc., so that bias and certain measurement errors will be minimized. The average tensile strength and standard error is

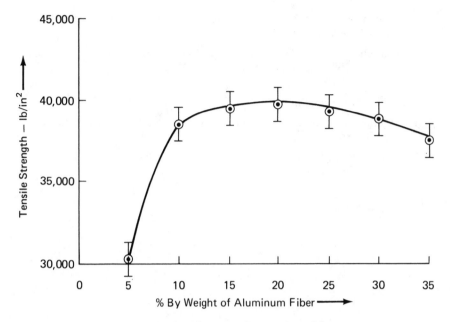

Fig. 7-10. Tensile strength vs. composition —
average data showing "confidence" limits.

calculated for each composition. The new curve of Fig. 7-10 can now
be drawn. The circles indicate the average tensile strength for each
composition; the "I" shaped figures show the upper and lower limits
of confidence as defined by the standard error at each point. Note
that the standard error may also vary with composition. Inspection
of Fig. 7-10 now shows that there is a definite improvement in epoxy
strength with the addition of aluminum fibers. The graph also indicates that adding more than 35% or less than 9% aluminum fiber is
not worthwhile. There appears to be a maximum effect somewhere
between 17 and 22%. It would, therefore, be valuable to explore this
concentration range more thoroughly.

References

Experimental Methods

* 1.   Strong, J., *Procedures in Experimental Physics,* Prentice Hall, New York
       (1938).

*Moderately technical.

* 2.   Wilson, E. Bright Jr., *An Introduction to Scientific Research,* McGraw Hill, New York (1952).

* 3.   Whitehead, T.N., *The Design and Use of Instruments and Accurate Mechanisms,* Macmillan, New York (1934).

Statistics and Data Handling

* 1.   Arking and Colton, *An Outline of Statistical Methods,* Barnes and Noble, New York.

** 2.   Fisher, R.A., *Statistical Methods for Research Workers,* Oliver and Boyd, Edinburgh (1932).

*Moderately technical.
**Advanced.

8

The Psychology
of Invention

Introduction

Inventive activity is, to a considerable degree, a mental process. It is not surprising, therefore, that various psychological factors will greatly influence the conduct of the innovative procedure. Inner needs, early influences, society's pressures, personal problems, and ambitions exert considerable force on the creative process. The amount of work done, the direction it takes, whether or not it is completed, the stage to which it is developed, and even the quality of the marketing effort will be subject to these forces. If the inventor can first evaluate some of his basic drives and subconscious influences, he can often avoid approaches or programs which will prove ultimately unprofitable. He can, subsequently, direct his efforts along more rewarding channels.

Motivation

Why invent? What is the basic and underlying drive for engaging in inventive activity? It is obviously a pleasurable activity for the participant or he would not normally pursue it. But what is the nature of the pleasure produced?

Inventive activity for some serves as a spare-time diversion, a hobby interest. As long as this type of inventor realizes the extent to which he is willing to become involved, he can adjust his aims and ambitions accordingly. Should he decide to devote full time to inventing or to the exploitation of the invention, he must take into account the corresponding changes which will be required in his lifestyle. Inventing as a full time activity involves many disappointments and setbacks. Years of routine measurements may yield only a small number of successful results. If he wishes to commercialize a single invention, on the other hand, most of his time will be spent in relatively mundane matters such as raw material selection, development of production techniques, writing of advertising literature, etc. Unlike the situation where inventing was merely a hobby, he may find the activity too demanding. Many aspects of it may become disappointing or boring.

Some individuals who invent as a hobby are inclined to "intellectual grazing"; they wander from area to area producing minor improvements first in one field and then in another. These persons cannot "invent to order" because they do not have sufficient attention span to carry through the many small steps required to perfect an idea and bring it to a marketable level. Because they do not or cannot control the subject matter of their inventions, they find it necessary to seek markets for what they have already devised. This considerably reduces their chances for success. A potential customer is much harder to convince if the concept, or at least a statement of the need, did not originate with him. A production superintendent of a tool company, for example, may assume a defensive stance if approached by an inventor who has perfected a better tool than his company now makes. If he accepts the invention and it fails to sell, he will be blamed. If it succeeds, he may feel that company management will credit only the inventor. The inventor's position is obviously much stronger if the superintendent realized a shortcoming in one of his products and commissioned the inventor to find a way around the problem.

The independent inventor is often not prepared to devote sufficient time to the brute force work of production or to a sales campaign. He seeks outside help to carry out these functions. Partly because of the risky nature of new developments and partly because of a shortage of development-minded entrepreneurs, the percentage of the profits demanded by the latter is usually very high. The end result is that the inventor of a highly successful product often barely realizes the equivalent of a salary had he been working directly for the entrepreneur. The unwillingness of many inventors to engage in the "crassly commercial" aspects of invention development and sale has also led to the springing up of fraudulent organizations which require the inventor to invest his own money and which then do a half-hearted (or no) job in promoting the invention.

It is well known that many inventors have, in the past, achieved wealth and status as a result of the commercial success of their innovations. It has usually been the case, however, that these individuals were talented businessmen as well, and were willing to forego the intellectual delights of continually exploring new and fascinating fields.

Let us now consider the individual who has examined his motives, finds that he primarily desires financial success, and is willing to devote a major portion of his time to the exploiting and selling phases of any promising invention he may at some future time develop. Should he now continue to invent at all? There are millions of unexploited inventions in the public domain from which he can legally choose. Expired patents and some patents owned by government agencies are available at no cost. Many currently protected ideas can also be used by the payment of relatively small royalties. While it is true that the individual following this course might not have the exclusive right to a particular idea that would be his if he had conceived and patented it himself, the item might be of a short-lived, high volume type so that exclusiveness would not be important. The fact that so many expired patents are not even considered for industrial exploitation indicates the existence of a powerful motivation in most inventor-entrepreneurs: the ego satisfaction in deriving success from one's own creativity. This personal characteristic is probably a major factor in the development of the inventive personality. It is important, therefore, that the inventor study his own motivations so that he may properly organize his resources. If he knows to what extent he really wishes to convert from tinkerer to scientist to businessman and in which area he desires to remain, he will have considerably bettered his chances for success and fulfillment.

Inner Needs

The choice of project, the approach taken, and the conclusions drawn from experiments — in short, a major part of the inventive process —

is dependent on the inner needs of the inventor. When emotional conflicts occur, the creative activities of the individual are among the most strongly affected of all the thought processes. The late psychologist Dr. Lawrence S. Kubie made studies of the influence of various neurotic tendencies on the creative process. He found interesting correlations between deep seated turmoil and the type of invention produced. An inventor of hearing aids, for example, may come from a background in which deafness was a significant factor in the daily life of his family. This individual finds it difficult to apply his skills to any other area because his activities are directed again and again toward alleviating a despised condition. Sometimes an inventor chooses his subject because of a residual hatred for some individual. Kubie mentions one medical researcher who spent years studying water metabolism. His subject, his approach, and his findings were biased by an early history of enuresis and the memories of the resulting humiliation caused him by his father. Another inventor so hated his father that all his innovations involved destructive devices and weapons (which had the potential for destroying the older man).

The influence of inner conflicts on how an invention is developed is illustrated by some inventors who insist on complete novelty. Not only does the development show a high degree of originality, but the measurement methods, the apparatus, the instruments, and the type of circuitry used are all unique. The work is made many times more difficult because each unique method must be experimentally developed before it can be trusted by the inventor. This type of behavior is sometimes found in persons who had basic difficulties with school subjects. Being unable to master standard techniques and fearing exposure, they compose their own methods for carrying out basic operations and tests. Another type of individual who exhibits extreme originality is the inventor who subconsciously rejects all authority.

Insecurity manifests itself in a number of ways in the inventive process. The inventor may maintain extreme secrecy, often justifying his behavior on the grounds that his ideas will be stolen. He is aware that witnessed and dated documents are necessary to assure claims of priority but is afraid to bare his work to public scrutiny and possible ridicule. This individual will sometimes work on interdependent projects as a way of delaying final solution and public evaluation. His primary invention, for example, may depend for its success on a new material which he must first develop. As a result, no project ever seems to reach fruition.

If the individual has deep fears that his employer will disapprove of his spare-time inventing activities, he sometimes finds that initially encouraging data cannot be repeated as the development reaches completion. Data is subconsciously or consciously altered as the point is approached where the inventor is about to make public his discovery. Sometimes a professional scientist will come upon a result which he fears will imperil his job; he may be tempted to delay or modify his findings in order to preserve his source of income. The former head of an environmental agency altered the reports of subordinates and discarded data which showed certain pollutants to be well tolerated by the body. He apparently felt that the continuation of appropriations to the agency depended on maintaining the idea that all contaminants are deadly.

Sometimes inner fears take the less-focused and less-defined form psychologists call "anxiety," which may arise from feelings of unworthiness or incompetence. The independent inventor often lives and moves in non-technical circles where he may be given somewhat unsympathetic treatment. In some anxious individuals there arises a need for a daily reward. As a result, they choose simple, easily completed projects. An inventor may be competent enough, for example, to do significant work in radar, but may elect instead to develop new types of kitchen utensils. In his frantic effort to put each day to a successful rest, he avoids projects with heavy technical involvements where victory is not so easily won.

Anxieties may seize an entire organization and result in a large-scale distortion of the creative process. Massive layoffs and greatly restricted budgets in the 1970s so affected the remaining workers in government and aerospace laboratories that a concerted effort was made to justify the continuation of the work. A "spin-off" program was set up to encourage industry to make use of space developments and thereby provide additional justification for expenditures of public money on rocket probe launchings. Although the spin-off program has led to considerable application of some of these developments by industry, the wide separation between the space and commercial environments in terms of overall objectives, cost effectiveness and methods used makes high profitability very unlikely. What is needed is a guaranteed and politically undisturbed annual appropriation for space research and an equally constant supply of funds for the development of inventions useful to industry.

Maturity

Maturity would not seem a requisite for successful creativity. The rigidness associated with age is not conducive to the imaginative flights which lead to brilliant new concepts. We have come to accept youthful exuberance as part of the inventive personality. There is, however, a distinction between youthfulness and immaturity. The latter often involves a desire to escape reality or avoid responsible involvement. The immature inventor may, for example, leave an unfinished project to take up another and then start a third without ever returning to the first. After some years, he has a workshop full of semi-completed models and experiments. Sometimes one of the ideas in his shop will appear as a finished product on public sale, the work of some more diligent inventor. The immature individual will often complain bitterly that his work has somehow been purloined and he has been cheated of the success which is rightfully his.

Immaturity will sometimes take the form of father image worship. Almost every well known scientist, inventor, or professor has disciples who walk in the shadow of his authority. The relationship may be that of teacher and student, employer and employee, researcher and assistant, or merely host and hanger-on, but it represents the same problem — the reluctance of the immature individual to make independent decisions. In graduate schools, there is often found the "professional student" who does investigative work under guidance of an eminent professor and contributes to a long list of papers which carry his mentor's name first. The student has long ago reached the point of educational saturation but he is unable to leave the university and do creative work on his own.

Thought Proccess

Although the presence of certain subconscious processes can distort or limit creativity, it is also possible to employ other subconscious mechanisms to *aid* in problem solving. In Chapter 4, we described a number of methods for creativity enhancement. Various free association techniques were discussed, whereby new combinations of ideas would be encouraged. By drawing at random from subconscious memories, it is also feasible to achieve solutions to problems. A well known and highly effective means for utilizing subconscious processes is to work on a problem before retiring. It will often be found that several possible solutions will occur on wakening. It is important that these solutions be written down immediately because they tend to be forgotten in a very short time after arising.

A phenomenon which can sometimes prove embarrassing is that of subliminal learning. A new device or solution to a problem is glimpsed on passing and the incident is forgotten. At a later time, the memory appears as an original concept. There has been no subconscious combination in this case; merely recall of a subliminally acquired concept.

It is often surprising to an inventor that a patent search uncovers a device almost identical to his own which was conceived and developed within almost the same period of time. It would not have been possible for the other inventor to have gained knowledge of the present work. It is tempting to attribute such occurrences to extrasensory perception or other esoteric thought processes. It is more likely, however, that a particular technology at any stage of its development will be of interest to a surprisingly large number of inventors. All of these individuals are well versed in the present state of the art and are familiar with its shortcomings. It is not really surprising that two inventors will arrive at the same solution at almost the same time.

In this chapter, we have described a few of the mental processes contributing to and hindering the act of invention. If the inventor is to achieve his goals, he must, of course, have a clear idea of what they are and how much will be required of him in time and work. If he has had considerable difficulty in producing useful improvements in his chosen area, there may be blocks which can or cannot be removed. If he desires to carry on further with the commercial exploitation of his inventions, he must again assess his goals, estimate the requirements, and determine if he is prepared to become sufficiently involved to bring about the desired result.

References

1. Kubie, Lawrence S., *Neurotic Disturbance of the Creative Process,* New York.

2. Kubie, Lawrence S., "Some Unsolved Problems of the Scientific Career," *American Scientist,* pp. 596-613, XLI (Oct. 1953).

9
Obtaining a Patent

Introduction

Somewhere between the conception of an idea and the completion of an operating model it may occur to an inventor that he needs the protection offered by a patent. In deciding whether to apply for a patent or not, he should ask himself two basic questions: (1) Is the device patentable? and (2) Will the obtaining of a patent have any advantage for me? An affirmative answer to both questions may justify the expense involved in seeking a patent.

The first question can be expanded into a number of individual items which define the overall area of patentability. As we saw in Chapter 1, a patent can be obtained on a process, machine, method of manufacture, or a composition of matter. Special patent protection is also available for designs and for trees, shrubs, fruit, and other plants produced by grafting, budding, etc. The discovery of a scientific principle, a method for doing business, or a naturally occurring article cannot be patented. If the invention appears, in the judgment of the Patent Office, to be an obvious solution to the problem at hand and could have occurred to anyone having ordinary skill in the art, it is not patentable. If the improvement has been made and patented by someone else, it cannot, of course, be patented by the later inventor. It is to minimize the possibility of anticipation by prior art that a search (Chapter 3) is highly desirable.

In addition to the above qualifications for a patent, there exist certain requirements which are based on acts by the inventor and others. If he publishes a description of the invention in a journal or newspaper or gives a lecture describing the device to an audience and then allows 12 months to elapse before filing his patent application, he has created what is termed a "bar" to patentability. It is assumed that the inventor has, under these circumstances, donated his findings to the public good. He is, therefore, no longer entitled to the exclusive use of the invention. The same bar is set up if anyone else having knowledge of the invention (either by learning of it from the inventor or discovering it independently) publishes a description and 12 months elapse before the application. Another act by the inventor which loses him his right to a patent is the filing of an application in a foreign country more than 12 months prior to his filing in this country and the granting of the foreign patent.

Illegal procedures by the inventor will also prevent patenting. If he attempts to exclude a co-inventor, for example, the Patent Office will deny him a patent.

Still another bar to patentability arises if the inventor manufactures the improvement, places it on the market, and sells it for a period exceeding 12 months prior to his application.

The various laws and regulations pertaining to patentability can be found in the government publications listed at the end of this chapter. These references are kept in many libraries or can be purchased from the Superintendent of Documents, U.S. Government Printing Office, Washington, D.C. 20402.

The second question — concerning the advantage to be gained by obtaining a patent in any particular case — can be answered in the affirmative if the situation meets one or more of the following conditions:

1. The inventor wishes to manufacture the improvement himself and desires to prevent anyone else from making, using, or selling it.

2. The inventor wishes to sell or license the improvement to a manufacturer. He wishes to present the idea as a saleable, legal entity and not merely as a concept which can be appropriated by anyone learning of it.

3. The inventor wishes to obtain employment with some organi-
    zation or to obtain a development contract. A complex patent
    is often of limited value to a company unless the services of
    the inventor are also available as a consultant or production
    engineer.

4. The inventor desires to build his reputation as an authority in
    certain areas. He is well funded and wishes to establish public
    knowledge of his expertise. By building up a list of patent
    holdings, he seeks to intellectually dominate a particular field
    or to profit by licensing his patents to manufacturers.

Another factor in considering whether a patent would be of advantage is the effect of not seeking protection at all. If the device will
likely be of the short-term, high volume, or "fad" type, then a patent
is really not required. Many toys, games, home exercise devices, and
novelty items appear briefly but enjoy a wide sale. In less than a
year, sales fall off. Competitors enter, offering lower prices, and
finally, volume and profit disappear. The originator has, by this time,
realized a maximum financial benefit and is already starting out on a
new development.

In some composition of matter inventions, it is possible to keep
the formulation secret. Would-be competitors must turn out a somewhat different composition, which becomes identified in the public's
mind as an inferior product. The Coca-Cola formula, for example,
was never patented. Had it been, the patent would have expired
years ago and legal duplicators would have swarmed into the field.
Secrecy is becoming more difficult to maintain because of the greater
capabilities of present day analytical equipment. It is often the case,
however, that the order in which certain steps are taken is as important as the amounts and types of ingredients used. It is difficult for
a competitor to determine the proper order by any method other
than the time consuming one of trial and error.

Patent Services

The patent system as originally provided for in the Constitution en-
visioned direct dealings between the Patent Office and the inventor.
It is still legal for an inventor to apply for and prosecute his own

patent; many thousands of inventors have done so in the past. It is also legal for a person to act as his own attorney in a court case and many have also done this. The arguments against the two practices are similar. Over a period of many years, legal niceties, interpretations, rulings based on precedents, special cases, etc., have introduced a considerable complexity into what once were fairly straightforward procedures. The inventor who nowadays chooses to apply for his own patent must, in addition to his other skills, be versant with the preparation of specifications, the writing of claims, the amendment of applications to meet examiner objections, appeals practices, and the details of interference procedure. Even if he acquires a working knowledge of these matters, he is still in the position of the "parent of the brainchild" and cannot bring to bear an outsider's impartiality — a detached view — which is often necessary for the most effective prosecution of the application. He may, under favorable circumstances, be able to obtain a patent, but its coverage or eventual validity may be impaired by this limitation. It is highly recommended, therefore, that the inventor resist the temptation to save money and that he employ outside help in seeking a patent.

There are several kinds of professional talent which he can employ. These are listed below:

1. Patent attorneys
2. Patent agents
3. Advisers
4. Development companies

A patent attorney is a member of the bar in the state in which he practices. He has also been accepted to practice before the U.S. Patent Office. He can handle all matters which come up between the inventor and the Patent Office, as well as court procedures which may or may not be required. The attorney is skilled in legal wording and can usually offer help with technical matters, such as the overall feasibility of the idea, violations of natural laws, etc.

A patent agent is basically trained in some technical field — chemistry, electrical engineering, physics, etc. — and has taken and passed the same test required of the attorney. Passing this test qualifies the individual to practice before the Patent Office. A patent agent can handle directly all matters which pertain to the Patent Office but cannot represent his client in court actions. He is competent in technical matters and can help the inventor select for patenting the most feasible embodiments of the development. Patent attorneys and agents are registered with the Patent Office.

An advisor will draw up patent applications for his client but these must be sent into the Patent Office by the inventor himself, without the name of the advisor. The latter does not generally help with the follow-up; the inventor must deal with such matters as claim rejections on his own. Advisors are not registered with the Patent Office.

Some development companies retain patent attorneys; others use the adviser system. In the latter case, the inventor must again deal with the Patent Office himself. It will generally be less expensive for an inventor to select his own patent lawyer or agent than to use those connected with a development company.

As mentioned above, a register is maintained of all patent attorneys and agents who are currently eligible to handle patent work for inventors. This register is kept in libraries and should be consulted to select a qualified individual who is in local practice.

Attorneys will be most effective in legal matters, agents in technical details. Fees charged by agents will tend to be less than those of attorneys, and since most patents do not involve court actions, an application can usually be adequately handled by a patent agent.

Costs

It is difficult to quote exact figures for fees and charges because the work required of an attorney or agent varies both with the complexity of the application and the follow-up required. As a rough estimate, the cost of preparing a patent application will range from $800 to $1,500. Office action responses which are required each time that the examiner finds objectionable features in the application will cost about $200 each. Added to this are Patent Office fees, drawing charges, and search costs. The Patent Office fees are as follows:

1. A filing fee of $65. This entitles the applicant to one independent claim and as many as nine dependent claims. An

additional charge of $10 is made for each independent claim in excess of one and up to ten. Claims in excess of ten cost $2 each.

2. An issuance fee. This is partially based on the "size" of the final patent. When the patent has been granted and is about to issue, the applicant must pay $100 plus $10 for each page of text and $2 for each sheet of drawings.

Drawings must be prepared by a competent draftsman who is knowledgeable of the techniques required by the Patent Office. Many inventors who have acquired skill in drafting have been able to make their own application drawings by following the conventions described in the book, "Rules of Practice," issued by the Government Printing Office. If it is necessary to have the drawings done by others, a minimum charge of $75 per sheet should be expected. The cost is higher when the subject matter is complex or when there are several figures per sheet. A preliminary search costs about $65; long and thorough searches are considerably more expensive. The inventor is charged $45 per hour of the searcher's time. All of the above costs are based on prices as they stand in the early 1980s and will likely change with time. A relatively simple patent having five pages, two independent claims, two sheets of drawings, and requiring one office action might be priced as follows:

Attorney's fee.	$800
Filing fee	75
Two drawings at $75 ea.	150
Search, 3 hours at $45 ea.	135
Response to first office action	200
Issuance fee, 5 pages at $10, 2 sheets of drawings at $2 + $100	154
	$1514

This is a relatively large expenditure and must be carefully weighed by the inventor against the estimated ultimate worth of the patented invention. The price of a patent can, in some respects, be compared to that of an injection mold. If the product is successful and many units are produced, the cost per unit is quite modest. If the patented invention is sold or licensed to another, the asking price will cover the cost of the patenting. If the product or idea does not sell, the money invested in the patent is largely wasted (as would also be the case with the injection mold). In medium and large scale industrial projects of a speculative nature, patent expenditures represent a small fraction of the capital outlay normally required for initial inventory, storage, advertising, etc.

Working with an Attorney or Agent

After engaging an attorney or agent, the inventor will be asked to sign an empowering form. The latter appoints the selected individual or firm to act for the inventor in matters dealing with the Patent Office. It is not permanent and may be withdrawn by the inventor at any time. The attorney or agent can, on his part, also withdraw from a particular case, but he must first obtain approval from the Commissioner of Patents.

When an attorney or agent is engaged, he should be given the complete trust and cooperation of the inventor in order that he can provide maximum assistance. Changes in scope or new discoveries should be revealed to the attorney or agent as soon as possible. Inventors will sometimes hold back a finding made after the application work has started because they fear imposing a delay on the procedure. This is extremely unwise and may result in the issuance of a weaker patent than would otherwise have been the case. Some inventors expect too much from an attorney or agent — unlimited consultation on other problems, marketing services, or the obtaining of special favors from the Patent Office. The attorney or agent must charge for his time and cannot offer free consultation. He does not generally engage in marketing services and he has no special influence with the Patent Office. Delays because of heavy backlogs will affect all applicants; no attorney or agent has the power to cause one application to take precedence over another.

Some inventors fear that revealing all details of their innovation to an attorney or agent will result in the compromise of the idea. It is, on the contrary, to the best interest of the attorney or agent to maintain a spotless reputation with regard to the confidential information revealed to him by a client. There is, first of all, the matter of professional and community goodwill. In addition, the Patent Office enforces a strict code which spells out the ethical requirements in order for an individual attorney or agent to remain on its registration list.

In helping to prepare the application, the inventor should write as complete a description of the invention as possible. It is he who has the most intimate understanding of the technical aspects of the improvement. The attorney or agent will modify the wording to incorporate it into a specification and will then abstract certain portions. From the latter, he will formulate one or more claims — single sentences which set forth the new features which the inventor wants to establish as his unique contribution to the art.

The actual documentation sent into the Patent Office is comprised of the following parts:

1. A petition to the Commissioner of Patents asking that a patent be granted on the accompanying improvement.

2. A signed oath or declaration stating that the inventor believes himself to be the true originator of the improvement, that it was not sold on the public market for a period exceeding 12 months prior to the application, that a patent was not obtained in a foreign country one year prior to the U.S. application, and that it was not described in a printed publication one year or more prior to the application.

3. A power of attorney appointing a specific attorney or agent to act for the inventor. All mail and telephone communications with the Patent Office will then be with the appointed individual.

4. The specification, drawings, and claims. A detailed description of these documents is given in Chapter 3.

5. The required fees.

The Patenting Procedure

A few weeks after the above package is submitted, the Patent Office will return an official receipt listing the serial number which has been assigned to the application. This number will subsequently identify the application in all future communications. The inventor has now reached the "patent applied for" stage. If he feels that no prior art exists which can prevent his patent from being granted, he may now begin to manufacture and sell the invention. A somewhat safer procedure, however, is to wait for the first response by the patent examiner assigned to his application. This response, called the "first official action," usually requires six months to one year. It may consist of one of the following:

1. Complete acceptance of the specifications, claims, drawings, and other documents.

2. Objections covering defects or shortcomings in one or more of the documents — e.g., incomplete declaration, insufficient fee, non-standard drawing, etc.

3. Objections to the specifications — e.g., the specifications do not describe exactly the same invention as do the claims, the specifications describe more than one invention, parts of the specifications are not clear, etc.

4. Rejection of some or all of the claims because the invention claimed has been anticipated by another inventor, the invention as claimed would be obvious to anyone reasonably skilled in the art, the claims encompass more than the specifications describe, the claims are too general and would tie up all future developments in this area, the claims are incorrectly constructed, etc.

5. Objections or rejections as listed above with specific recommendations for modifying the application. The modifications would then place the application into a form acceptable to the examiner.

An application is seldom accepted as submitted. The examiner, skilled in search and evaluation and with all the facilities of the Patent Office available to him, can very often unearth prior art that has escaped even the most thorough search. Although the requirements imposed by the examiner may seem rigid, it is his monitoring which keeps the tremendous number of patents legally separate and prevents a totally hopeless tie-up of individual proprietary rights.

Included with the first official action is a time requirement. The attorney or agent must respond to every objection within the set time period — usually six months, but this can be shortened to as little as 30 days at the option of the examiner. If response is not made within the required time, the application is considered abandoned and all further consideration by the Patent Office stops. Under special circumstances it is possible to revive an abandoned application. Time for revival is also limited.

The response by the attorney or agent in the case of defective specifications or other documents is to modify them in conformance with the examiner's requirements. Defective claims are amended by appending or deleting sentence structures. No new material is allowable — i.e., material which expands the original scope of the specifications, or additional claims.

After the response has been received, the examiner judges whether his objections have been met. If there are still objections, he will return a second official office action. Some claims which were originally rejected may again be considered unacceptable. The inventor's side now has one more chance to modify the documents for acceptance. If this "final" try is not successful and the examiner again rejects the claims, the inventor's attorney or agent may appeal to a Board of Appeals. The latter consists of the Commissioner of Patents, the Assistant Commissioner, and usually, three examiners-in-chief. If this board rules for the inventor, the patent will be granted. If a decision upholding the original examiner is obtained, appellate action may be extended to the Court of Customs and Patent Appeals or a civil suit may be filed against the Commissioner of Patents in the United States District Court of the District of Columbia.

If the application is approved after an official action, the applicant's attorney or agent receives a notice of allowance and is billed for final fee. When the latter is paid, the patent is printed and a copy of the patent and the certificate of ownership are mailed to the attorney or agent.

Although the process appears complex and difficult to carry through, the chances for an inventor to obtain a patent are very good. Patents are granted on approximately 65% of the applications submitted.

Before we conclude this chapter, several lesser-known patent forms will be described. As mentioned above, once a patent application has been filed, new material may not be added. Sometimes it is necessary to correct a mistake in the original application or even in an issued patent. An application can be made for a "reissue" patent which adds corrective matter.

New matter may also be added to the patent prosecution by filing another kind of second application; this seeks to obtain a "continuation" patent. The latter can be used to expand the original concept in terms of new findings. In many cases, the original application is allowed to become abandoned after the continuation application is filed.

If the examiner finds that two or more distinct inventions are represented in the application, he may require the inventor to elect one invention and restrict the claims to that one alone. The non-elected inventions can subsequently be protected by application for one or more "divisional" patents.

If reference to the original application is made in the reissue, continuation, or divisional application, the inventor retains the priority of the early date shown in the original.

References

Publications of the U.S. Patent Office; obtainable from the Superintendent of Documents, U.S. Government Printing Office, Washington, D.C. 20402:

1. *Patent Laws*
2. *Rules of Practice*
3. *Manual of Patent Examining Procedure*
4. *Patents and Inventions. An Information Aid for Inventors*
5. *General Information Concerning Patents*
6. *Guide for Patent Draftsmen*

General Books of Interest

7. Tuska, C.D., *An Introduction to Patents for Inventors and Engineers,* Dover Publications, New York (1964).
8. Fenner, T.W. and J.L. Everett, *Inventor's Handbook,* Chemical Publishing Company, New York (1969).

10

Making Your Invention Pay

Basic Considerations

Marketing is often an activity foreign to the inventor, or at least not a preferred form of work. A man who is outstanding at finding solutions to technical problems is frequently ill-equipped for the selling end of invention development. Selling can, however, be as much a challenge to creativity as is the inventive process itself.

A block to effective marketing is often set up by a lack of goal definition. It is important that objectives be stated as early as possible and likely paths for reaching these goals be listed.

The inventor might begin by asking himself: "What do I want from this invention? What is the minimum amount of money I would need to pay my development costs and then to provide a fair profit for the work I have done? What is the maximum amount I could hope to get under the very best of circumstances?" Answers to these questions will set a range for his asking price. Perhaps money is not really the prime consideration in the case of a particular invention — the inventor may wish to gain contacts in some industry for which he will then do more work at a later time. Or the inventor may only desire to put to use one or more promising devices which have been gathering dust on a back shelf for a long time.

Another important question to be asked is, "How much do I believe in this project?" Unless he is firmly convinced that the invention possesses great utility and strong advantages over existing art, the inventor will have difficulty convincing a prospective buyer.

It is difficult to assay the importance or potential importance of any invention, especially if it is one's own creation. When the paper clip was devised, it might have appeared to many as an insignificant item, yet its wide use has certainly established it as a development having a high degree of importance. By contrast, a golf tee made of cardboard was invented a few years ago. It supported a golf ball without the need for forcing a sharpened object into the ground, did not break easily and was inexpensive to produce. The invention was not acceptable to manufacturers because it seemed too insignificant an improvement. The inventor did not have sufficient confidence to produce the tee himself and gamble on the public's acceptance. Inventions at the other end of the spectrum of complexity present similar problems when importance is being judged. The invention may, for example, relate to an analyzer for determining the percentage of chromium in steel by a spectrographic process. Although millions of tons of steel containing chromium are produced annually, they can be analyzed by relatively few instruments. Existing devices are well established and it would take a considerable amount of convincing to sell either the instrument manufacturers or the steel companies on this invention.

Exploring the Possibilities

The simplest approach to converting an invention into money is to sell an unpatented idea to a manufacturer. Many leading companies will look at an unpatented idea if the inventor will first sign a "disclosure form." This much-argued-about document limits the obligations of the company. The invention will be treated on a good faith basis by most companies but no guarantee of confidentiality can be given. Because it is possible that someone within the organization is already working on a similar development, the company does not bind itself to a "pay or not use" policy. In many disclosure forms, the maximum payment for an unpatented invention is set at $1,000. In view of these limitations, it might seem undesirable for an inventor to deal on these terms; it must be remembered, on the other hand, that one or two initial sales will establish an inventor's reputation with a company and provide him personal contacts within the organization. The inventor can also learn what kind of problems the company is really interested in solving. Subsequent inventions can be negotiated under much more favorable terms, because the company will then know the individual's capabilities and be interested in providing the maximum incentive for future work.

The inventor who deals on the disclosure form basis should not send a "bare bones" idea to the manufacturer. An idea's salability is greatly enhanced if the inventor has tested it, can provide performance data, or can demonstrate a working model. If a patent search has been made, the findings should be included. Since a negotiation for an unprotected idea is conducted on an "arms length" basis, the inventor would do well to retain to himself some aspect of the invention. Should the company be somewhat less than ethical and his idea is appropriated, he will still have a portion of his innovation left for possible sale to another company or for a patent application. In the description he gives the company, the inventor should reveal "how" something is done but should never state "why." If he reveals his underlying investigative philosophy, others can extrapolate and can arrive at the same improvements he will later achieve. An illustration of the processes involved in the sale of an unpatented idea to a manufacturer is given in the following hypothetical example.

Let us assume that the invention comprises an "eraser key" for a typewriter. When struck, this key deactivates the ribbon mechanism and substitutes a portion of a "white carbon" strip. The latter is stored between two miniature reels located just under the regular ribbon. The striking head of the eraser is a flat-faced metal rectangle of sufficient size to cover any letter or number. Upon impact, the striking head deposits a white, rectangular imprint on the paper to cover and hide any incorrect character. The imprint can now be typed over with little or no trace of the previous entry. The inventor builds a working model and installs it on a typewriter. He establishes its workability and utility to his own satisfaction. He then lends the typewriter to a number of public stenographers who find it a useful improvement and sign statements to that effect. Because the invention is an addition to a larger machine and must be adapted for each make of typewriter, the inventor feels it would be better to sell the idea, rather than attempt to manufacture it himself. Correspondence with a large producer of business machines leads to the signing of a disclosure form. The inventor then submits a document showing the scope of the invention, the results of a search he made, including copies of patents which came closest to his idea, the signed statements of the stenographers, and some supplemental ideas on how the white carbon ribbon could be loaded into the machine. During the inventor's original experiments, he found that imparting transverse vibration to the striking head of the eraser significantly improved the transfer of the white material to the paper. He has held back this feature of the invention, prepared − if the company appropriates his idea − to use this as a further improvement to offer a second company, or to apply for a patent. The company likes the concept as presented and agrees to buy the inventor's rights for $2,000. It also agrees to pay him $200 a day for consultation as required. The inventor can now reveal the additional information to his customer as part of his consulting services. The company now wishes to apply for several patents, including the vibrator, and gives the inventor a three months' contract for consulting services. Further work by the inventor results in the development of a free-swinging weight which imparts vibration to the erasing head when the key is struck. A working relationship with the company has now been firmly established. The inventor finds it easier to sell the company other ideas. He later develops a hollow head eraser structure through which a fine white powder is dispensed. The white carbon mechanism is no longer required. The manufacturer now pays the inventor a royalty for the use of this and subsequent inventions.

The Patented Idea

When an invention has been patented, a somewhat different set of factors comes into play. Because the patent defines the rights of the inventor, it does not matter if the company has a previous interest in the subject of the invention or has an ongoing research program along similar lines. The inventor can sue for infringement if the company appropriates his idea. In a sense, the patent also limits the obligations of the company, which is free to use whatever is not defined by the patent. Examination of the latter by company personnel does not obligate the company to buy nor does it hinder company application for patents of its own.

A few companies look only at issued patents; most, however, will consider inventions in the patent pending stage.

A number of variables influence the respective positions of the patent holder and the company. If the patent is pending, the company reviewing it has no guarantee it will be granted; this may reduce the amount they are willing to offer. Issued patents, which, in the judgment of the company's attorney, can be attacked as to validity, represent a weak position of the patent holder. The company might decide to appropriate the idea and depend on invalidation to win an infringement suit.

A granted patent is a legal entity. Because of this, its use and transfer is governed by the rules which apply to other legal entities such as real estate, machinery, certain securities, vehicles, etc. The patent may be sold outright for a specified amount of cash, stock or other valuable consideration; it may be licensed (essentially rented out) to one or more manufacturers, or it may be sold under a royalty arrangement. In the latter case, there is a cash down payment and a rental fee based on the number of units sold.

The outright sale of a pending patent is a relatively straightforward procedure. After receiving the agreed-upon sum, the inventor submits a certificate of assignment to the Patent Office, along with a fee of $20. When the patent is issued, the name of the purchaser appears on the document as the assignee of record. The amount of money to be charged by the inventor for assigning his rights must take into account several factors. In view of the one time nature of the transaction, the price asked should be relatively high but not unrealistically so. The inventor should, if possible, shop around for several offers. If this cannot be done, he should estimate the time, effort, and money expended on the invention and then estimate what would be a reasonable return on his investment. Many profitable sales have been lost by inventors who have demanded too high a selling price. Although the large company is generally considered to have a "business is business" attitude and will try to purchase the invention at the lowest possible price, the inventor should also consider the other side of the coin. A new product represents a significant risk for the manufacturer. The latter must spend additional money for auxiliary development, tooling, manufacturing, distributing, and advertising. There is no guarantee that the item will be a profitable addition to his line of merchandise. The inventor who is able to work out a compromise between what his pride of accomplishment leads him to believe is an adequate reward and what his intellect tells him is reasonable from the manufacturer's standpoint is most likely to make a profitable sale. It has been one of the major aims of this book to instruct the inventor in techniques for greatly increasing his output. An inventor who sells many of his ideas can afford to price each more reasonably than can the individual who has only "one great idea" to show for many years of work.

The problem of pricing encountered in an outright sale is alleviated somewhat when a "down payment plus royalty" arrangement is made. In this case, the inventor is offering to share the risk with the manufacturer. If sales turn out to be low, so will the royalties. If, on the other hand, public acceptance is beyond expectations, the inventor will share in the extra benefits. The down payment is generally considered an advance against royalties and will be deducted from the latter. Often the amount of the down payment is minimal — between $500 and $2,000. This is especially true if further development of the invention by the purchaser is required. A method for compensating the inventor sometimes used in these cases is for him to be hired as a consultant by the manufacturer to aid in the development work. Royalties generally range between 2 and 7% of the manufacturer's selling price, with exceptions in special cases. A novelty or toy selling at 29¢ might only pay the inventor .5 to 1% royalty, but an extensive volume would bring him a large income. A highly specialized machine tool selling at $50,000, but having a very limited volume, would require a royalty of 10% if the inventor is to be fairly rewarded. In setting up a down payment and royalty arrangement, there are a number of mutual understandings (besides amounts of money) which must be settled between the inventor and the buyer. The most important of these are:

1. *The performance clause.* The buyer agrees to produce a minimum number of units per year, or at least to pay the inventor his royalty for a minimum number of units. Without this clause there is nothing to prevent the manufacturer from "shelving" the patent indefinitely so that the inventor receives only the advance paid when the agreement was signed. The

manufacturer may have wanted the patent to keep the item off the market (because of potential competition with a new device he was developing). Or the manufacturer may encounter unexpected expenses or budget cuts and decide to drop the project.

2. *Exclusive or non-exclusive rights.* The manufacturer who desires an exclusive license should be willing to pay a higher royalty for this privilege.

3. *Restrictions.* A license can be restricted in a number of ways. The manufacturer may be limited to offering the device for sale in a certain geographic area or within a certain state. The sale may be restricted to a particular application. If the invention, for example, pertains to portable electric generators, the license may restrict one manufacturer to producing small units for use in camps, on pleasure boats, etc. Larger generators for outdoor lighting or construction jobs would be licensed to another manufacturer. The advantage to a manufacturer of a restricted license would be a lower royalty without competition in his real area of interest.

4. *Sub-licenses.* It is often desirable or necessary for a manufacturer to have an affiliate company produce and sell the patented item. Unless this possibility is specifically recognized and treated in the agreement, it is possible for the inventor to lose royalties because the sub-license company is under no obligation to pay him.

5. *Auditing.* The agreement should stipulate that a representative of the inventor is entitled at regular intervals to inspect the sales records of the company pertaining to the item and to determine if the royalty payments are accurate.

6. *Infringement.* Although most companies have an adequate legal staff and will agree to defend the purchased patent against infringement or attacks on validity, this responsibility should be stipulated in the agreement.

7. *Bankruptcy.* The purchasing company may encounter financial difficulties which result in bankruptcy or other insolvency, and the agreement should state how the purchased patent

rights will be handled. The inventor should seek to have these rights returned to him for sale to another company.

8. *Right to terminate.* If the inventor should become dissatisfied with the company's handling of his invention or the company no longer wishes to pursue the manufacture of it, financial adjustments to cover these possibilities must be made, and included in the agreement.

Locating Buyers

The classic method of locating potential buyers is to consult a directory. The names of companies having a possible interest in the invention are obtained from the listings under the general area served by the invention — e.g., manufacturers of elevators or industrial cleaning compounds. The directory may be no more comprehensive than the yellow pages of the telephone book or it may be a multiple volume and comprehensive work like the Thomas Register or MacRae's Bluebook. The latter two directories are particularly rich sources of information, in that they not only list manufacturers in every state but also provide a rating based on their size and financial status and give the names of executive officers. For the inventor to obtain the maximum utility from any directory, it is important that he survey all the listing possibilities. If the invention were, for instance, concerned with fireproof gloves for the handling of hot plate glass, the inventor would need to check companies listed under "Apparel," "Gloves," "Safety Products," "Glass Manufacturing Supplies," etc.

A useful publication found in most libraries is the "Guide to American Directories." The latter is an alphabetically arranged compendium of directories which apply to particular industries. Thus directories are available for specific fields such as advertising, agriculture, air conditioning, amusement parks, arms, art supplies, etc. These directories represent a more concentrated source of possible buyers than any of the more general listings.

In selecting companies from a directory, the inventor should consider three areas of possible utilization of his item:

1. Those companies equipped to manufacture his invention by virtue of their other lines of production.

2. Those who buy related items from manufacturers and supply them to wholesale or retail outlets.

3. Those who incorporate smaller items into a larger product.

If, for example, the inventor has an improved kind of brake lining which stands up well under heat, he would seek the manufacturers of automotive equipment, in the first category; the wholesale distributors of parts to stores and garages, in the second; and the manufacturers of automobiles, in the third.

In preparing and using lists of companies, the inventor must keep in mind that he is using the "shotgun" approach. Out of 500 letters of inquiry there may be only 30 responses. Half of these may merely be polite answers saying that there is no interest on the part of the company. Following up the 15 positive responses, he may be able only to generate face-to-face interviews with three companies. If no buying interest develops, the inventor must now prepare a second group of 500 letters. To override the feeling of hopelessness which the shotgun process sometimes produces, the inventor must always keep in mind that he need find only one buyer.

A more direct approach than mass mailing is to advertise in various journals and newspapers. Those who respond will at least have some initial interest. The Patent Gazette devotes a portion of each issue to a list of patents for sale. For a cost of $3.00, an inventor can offer a patented item in this list.

A very effective advertising technique is to submit a one or two paragraph write-up and a professional-quality photograph of the invention to a trade magazine. Many of these periodicals contain a section entitled "New Products." The publicity (which costs the inventor nothing) often results in several hundred inquiries from interested individuals and companies. A call or visit to any good library will provide the inventor with the names of the trade journals most suitable for his particular item.

A paid advertisement is relatively expensive but is probably the most generally efficient means for locating buyers. The wording of advertisements is in itself an art. Professional help is available at no additional charge from the publisher of the newspaper or magazine used. The ad should be slanted so as to attract direct buyers, not brokers or the idly curious.

Classified advertisements in the Sunday editions of large metropolitan newspapers are surprisingly effective. Perhaps the reason for this is that the businessman who happens to see the ad is in a relaxed, "Sunday afternoon" mood, but is unable to entirely take his mind off business-related matters. Advertising in top financial newspapers and magazines will frequently draw the attention of individuals who control company policies and can therefore become very good prospects.

Another method for finding buyers is to exhibit a model at a trade show. The latter range from exhibits dealing with small business enterprises to special shows of new inventions. To obtain schedules and fees for these shows, the inventor should write to his state or city chamber of commerce.

In any campaign to find a buyer, the inventor must be prepared to send along a well composed follow-up to those who inquire. He should have available carefully worded and adequately printed literature describing the invention, its operation, its applications, advantages, estimated sales potential, etc. A brochure showing the results of experimental work, estimated manufacturing costs, and a listing of present devices with which the invention would be competing would also be valuable. The literature should obey a cardinal rule of all marketing — it should suggest a definite action by the prospective buyer. In short, it should "ask for the sale."

Patent Development Companies and Brokers

The inventor who is seeking a buyer will inevitably come across the patent development company or the patent broker. Much financial loss and overall disillusionment with the patent system has come as a result of relations with some of these organizations. It will be valuable at this point to discuss the general subject of patent selling through the development company or patent broker.

All of the creative arts — acting, writing, fashion designing, modeling, painting, etc. — are characterized by the difficulty that the beginner has in breaking into the charmed circle of the money-making practitioners. As a result of the obstacles encountered by the newcomer, it has been very easy for unethical promoters in each of these areas to bilk aspiring hopefuls by offering an easier path to success. The courts have in recent times tried cases against acting schools which promised to get their graduates into movies or television, portrait photographers masquerading as agents of Broadway producers, art schools who implied that each student's work was appraised by a nationally famous artist, and subsidy publishers who made little effort to distribute the paid-for volumes of their writer-customers. The art of inventing has not been immune from this sort of exploitation. Many invention development companies have, as their only source of income, payments from the inventors themselves. A few of these organizations make a token effort to sell an invention, but even then, the quality of service rendered is much lower than the inventor would be capable of performing himself. The advertising of these companies, developed over years of dealing with the proud creators of new products, resembles the sales messages used by child movie star agencies to appeal so strongly to stage-struck mothers. The literature and sales talk paint alluring pictures of the sudden wealth to be expected when an invention is sold. If the inventor could see the sales record of many of these companies, he would be appalled to find that over a period of years only two or three patents had actually been marketed!

Fortunately, it is relatively simple to detect a less-than-scrupulous development company. In order to engage their services, the inventor is asked to pay in advance for one or more of the following "services:"

1. To conduct a search of prior art. The quality of this search is generally very low.

2. To prepare drawings, material lists, costs, or sales literature for submission to manufacturers.

3. To prepare photographs or demonstration models.

4. To start an advertising campaign.

This kind of development company also demands a commission in the unlikely case that the invention is sold. This is guaranteed in an agreement signed by the inventor when he pays his fee and may run as high as 25% of the selling price.

Some development companies merely make photocopies of the inventor's papers and send them to companies selected from a directory. The inventor will pay $300 to $500 for a 50-company mailing. Other development companies publish a magazine containing a brief paragraph describing each of their current offerings. The magazine is distributed free to many industrial concerns. This service can cost the inventor as much as $1,800. Many manufacturers have such a poor regard for the development companies who send them this unsolicited mail that they automatically discard all such material. In these cases, the inventor is actually paying a considerable sum in order to damage his chances for a sale!

As has been the case in other arts, inventors have banded together to form mutual aid groups that meet to evaluate each other's developments, to trade ideas on the best methods for marketing, to socialize, and, in general, to maintain good morale among members. These groups often sponsor shows in which members' inventions can be displayed for a reasonable fee. Well-run groups can attract special consideration from large manufacturers, the chamber of commerce, local patent attorneys and agents, model makers, and business advisers. Some groups maintain shops or laboratories where members may do experimental work. Because the terms "development," "research," "engineering," "consultants," etc. appear in the titles of both the exploitative organizations previously described and in the names of bona fide inventors' clubs, a workable criterion is needed by the individual when he first contacts a particular group. The following "rules of thumb" are suggested:

1. The payment required should be based on an annual membership, not on each invention submitted.
2. The annual fee must not exceed $75.
3. Extra fees for the use of other services must be nominal; e.g.:
   a. Workshop use: $1.00 per hour or just the cost of the materials and electricity used.
   b. Exhibit space: $25 to $50 for a small table for the duration of the show.
   c. Consultation offered: club patent service, search, drawings, model making, etc. A discount of 10 to 40% should be obtainable. All prices should be compared with those of independent providers of similar services.

4. The organization should hold frequent meetings which all members are entitled to attend.

The remarks made above concerning development companies also apply to patent brokers. If one of the latter asks for a large down payment by the inventor, it is highly likely that this is his major source of income. An individual who is running a legitimate patent marketing business will operate solely on a commission basis: 10 to 15% of the selling price or royalty. He will also be highly selective in the inventions he undertakes to sell.

In concluding this section, we cannot over-emphasize the desirability of the inventor's acquiring a "let the buyer beware" attitude. Any development company or broker considered should be requested to supply at least four notarized letters from inventors whom they have successfully served in the previous year.

Manufacturing the Invention

The best method for making some inventions pay off is for the inventor to manufacture them himself. If he is willing to put in the hard work and undergo the well known risks in order to establish an independent enterprise, he may be in a position to enjoy the many rewards: satisfaction in achievement, higher-than-ordinary royalty, salary as an employee of his own firm, a yearly profit, cash appreciation of the physical plant as business increases, capital gain from stock he issues to himself, higher standing in the community as a creator of jobs, etc. The problems accompanying success are also many: need for meeting a weekly payroll, numerous taxes and licenses to administrate and to pay, employee welfare requirements to fulfill, need for complex accounting procedures, payment of creditors, the threat of competition, and so on. If the inventor possesses a psychological makeup which convinces him that the rewards outweigh the problems, he can probably make a success of manufacturing his own invention.

In setting up a system for producing and marketing the item, a number of operational alternatives are available to the inventor. These entail varying degrees of involvement, ranging from an almost "remote control" operation to one requiring complete participation.

The "lower involvement" types appeal to inventors who desire only to work up an operation, sell out at a profit, and then return to inventing. Below are listed a few of the possible procedural combinations in the order of the increasing participation required.

1. Production subcontracted to a specialty manufacturer. These organizations are set up to produce small and medium size runs of a wide range of items. Inventory is shipped to a warehouse; sales are handled by franchise holders who order through the inventor; merchandise is drop shipped from the warehouse. The specialty manufacturer may be located in a low cost labor market in the United States or in a foreign country.

2. Production by a manufacturer who makes a similar product for wholesaling. Item is placed on consignment in retail outlets by the inventor-entrepreneur.

3. Batch manufacture by the inventor in a garage-type operation. Inventory is stored on his own premises; advertising in magazines generates sales; shipping is done by the inventor and one or two assistants, often family members.

4. Custom construction of a complex item. No inventory except for one or two floor models or demonstrators. The entrepreneur receives and fills orders sent into his shop by salesmen.

5. Medium scale production in the entrepreneur's own plant. Marketing directly to customers by salesmen. Inventory maintained on the premises.

Many combinations of the above are possible, as well as special situations which are generated by the type of invention involved. In isolated instances, for example, it might never be necessary for the inventor to construct more than a few copies of his device. He might sell, as a service, the work performed by a skilled individual using the invention. A case of this type would be a new style of analyzer for diagnosing automobile systems. The inventor would set up a series of shops, each containing an analyzer, and charge motorists for the service of locating the sources of poor performance.

The size and type of the potential market will often dictate the best kind of manufacturing and distributing operation. If the item is a simple plastic molding which is to be sold in large quantities, the major investment by the entrepreneur in production equipment might be an injection mold. The mold is loaned to a plastics fabricator whenever it is necessary to rebuild the inventory. The finished item might be packed in cardboard boxes for shipment to wholesale outlets, or individual units might be enclosed in plastic "bubbles" joined to printed cards for point-of-display sale. If the product is an electronic sub-assembly, the most efficient production method might involve the purchasing of custom-etched circuit boards from a specialty house and the installation of components by the entrepreneur's people.

Another factor which determines the type of participation required of the inventor is the final selling price. The well known relationship between price and the number of units sold is generally valid. The higher the price, the greater will be the resistance to buying. If the newcomer is too expensive, it has less chance of displacing established competitive items. It is, of course, one objective of inventing to produce products of a unique nature whose utility cannot be matched by existing items. In addition, clever advertising and promotion can be used to build up the prestige of a new development and thus permit an unusually high selling price. A well known example involved an invention which retailed for 60 times the cost of the existing competitive device but afforded no improvement in performance — yet sales soon mounted to millions of dollars per year. The invention was the electric razor, which sold for $15 as compared to the 25 cent safety razor of the time. It is easy for the inventor to estimate his working parameters once he has arrived at a realistic selling price. If he is to wholesale the item, the selling price is cut in half and then further reduced by 10% of retail as a discount for cash. This will give him his expected gross per unit. Assume that the product will have a retail price of $8.00; the wholesale price will then be $4.00. There would be a trade discount of $.80 for cash payment within 10 days. Thus, the adjusted gross is now $3.20. The entrepreneur's choice of components, manufacturing techniques, and labor must then be adjusted so that he may work within this amount and still show a profit. If he can supply the item for $1.00, he will now have $2.30 to draw from for profit, shipping costs, unsold merchandise, breakage, rejects, etc.

In manufacturing his invention, the inventor-entrepreneur will need to have some knowledge of many job disciplines, including industrial engineering, raw material selection, business management, marketing, personnel selection, pricing structure, design, product styling, inventory control – just to name a few. The total knowledge to be acquired and applied, when coupled with that needed for inventing a successful product, may appear to be a staggering load; yet many inventors have met this challenge and have established huge and immensely profitable enterprises.

Miscellaneous Considerations

"Thinking big" has often been proposed by popularly read psychologists as a sure road to success. Raising one's sights is recommended as a method of rising above small ambitions and going onto large accomplishments. In the development and marketing of inventions, however, thinking big can sometimes be a disadvantage. Let us assume that the invention is a solar absorber that is intended to be placed in stationary orbit around the earth. The machine converts solar energy into laser radiation that is then beamed to a receiving station on earth. The independent inventor of this system faces some obvious problems. He may have demonstrated initial feasibility by a scale model but must convince the government or a group of very large companies to risk vast sums for a more realistic evaluation. The number of codevelopments necessary to make his invention operable on a practical scale (a space shuttle to maintain a steady flow of materials from earth during plant construction, space-worthy laser-generating machinery, efficient beaming and receiving antennas, etc.) would considerably dilute the value of his contribution. Even if he were successful in selling the rights to the invention, there might be a long delay in the payment of royalties because of the time required for its profitability to be established. He might thus fare worse than another inventor with more modest leanings. An additional, and unfortunate, consequence might be that his invention is turned down but a satellite power station of this type is eventually built. The completed unit would have been the result of the concerted effort of thousands of scientists and engineers. The required and eventually achieved degree of sophistication is orders of magnitude above that of the apparatus described in his patent. The concepts finally used were not really copied from his patent but came as logical steps in the space development program. The inventor nevertheless feels that he has been cheated.

This problem with complex inventions can sometimes be avoided if the inventor approaches government agencies with a research proposal. If interested, the agency will contract with the inventor to develop a more refined model or to study what difficulties may be encountered in actual practice. If a laboratory is required, the agency may arrange for his use of government-owned facilities. The inventor would then play the role of contributor to the efforts of a large number of others rather than attempting to compete with them.

The above statements are not intended to imply that the inventor should banish any creative thoughts concerned with exotic inventions or aim solely for easily achieved targets. He should, however, carefully preselect from a list of his ideas those on which he will spend considerable time and money in developing, patenting, and selling.

Another extreme in thinking that may cause problems for the inventor is that of too narrow a scope. Suppose a professional astronomer, for example, devises an instrument for improving the accuracy of dial readings. The instrument makes possible the calculation of interstellar distances to a degree not possible before. The invention is somewhat complicated and costly and cannot be improved in this respect because mass production is not feasible. Potential buyers of the invention would be that very small percentage of the population who are professional astronomers. The device will not work with an amateur's telescope because the dials of his equipment are too coarsely calibrated. Also, the amateur's interests are usually in the nonquantitative aspects of astronomy. Among professional astronomers there are a large number who now work with radio telescopes. These telescopes do not resemble the optical type and do not employ the particular kind of mechanism to which the invention applies. Among professional optical astronomers there are those engaged in theoretical work; these individuals make no measurements at all. Finally, the invention might only be purchased by a few observatories where it would be used by several professionals. The number of optical observatories is relatively constant. Once the device has been sold to all of these there is very little further demand. The invention, though unique and useful, thus has limited economic potential.

A factor often overlooked by the inventor is whether his improvement will stand alone or must be used as part of another invention. If the latter is true, can the improvement be easily attached by the user or must it be incorporated into the larger system at the factory? If the inventor wanted to manufacture the item, could he obtain the cooperation of the original patent holder? For a favorable situation in this regard, let us assume that the invention is a new type of flashlight switch that is immune to corrosion. This device must be built into the flashlight case, but the inventor is free to license any manufacturer who is interested in the switch because all flashlight patents are by now in the public domain. Assume on the other hand that the improvement is an automatic shutter control for a certain type of camera. Two rapid flashes are used. The amount of light reflected from the subject as a result of the first flash sets the shutter speed. The second flash, synchronized with the opening of the shutter, then serves for the conventional purpose of picture taking. This invention must be incorporated into a highly complex feedback system that is a patented portion of the presently manufactured camera. The inventor in this case must deal with the holder of the patent rights to the original invention; this might not be a favorable position for him.

A third situation could arise in the case of a new smog control valve for automobiles. Installation is simple and can be done by the average motorist. In this case the inventor is free to select his sales method. He can choose to produce the device with the full protection of his patent. He can also lease rights in it to a manufacturer of car accessories or to a maker of automobiles.

A final consideration in the evaluation of an invention's potential market involves consumable products and replacement supplies. It is sometimes found that the profit from an invention is dwarfed by the sale of spare parts or of materials to be used in its operation. Compositions of matter used with certain inventions often are consumed at a high rate and must be replaced. A cat-food dispenser, for example, may be a valuable patented item because of the steady consumption of specially shaped food pellets. As another example, the invention might describe an insecticide sprayer to be used with a certain kind of liquid. Sprayer sales might bring a modest profit but the sale of the insecticide itself would represent a large repeat business.

References

The following books contain much valuable information on selling patents, manufacturing one's inventions, setting up businesses, and financing.

1. Paige, Richard E., *Complete Guide to Making Money with Your Ideas and Inventions,* Prentice Hall, New Jersey (1973).

2. Fenner, T.W. and James Everett, *Inventor's Handbook,* Chemical Publishing Company, New York (1969).

3. Berle, Alf K., *Inventions, Patents and Their Management,* Van Nostrand Reinhold Co., New York (1959).

4. McNair, E.P. and J.E. Schwenk, *How to Become a Successful Inventor,* Hastings House, New York (1973).

11

Invention into Enterprise – Some Case Histories

The steps involved in making the transition between a workable invention and an active business have been previously outlined. To further illustrate the general method, a few examples will be presented that describe how different inventors under differing circumstances converted their ideas into profitable enterprises. The inventions chosen represent four widely separated areas of interest: an industrial product, an instructional method, a recreational device, and a medical aid. The companies vary from home-operated, part-time businesses, to small corporations. The rate of growth and final form of each enterprise depended on the business skills of the inventor and his associates, the nature of the invention, the condition of the economy into which the business was launched, and the luck of the entrepreneur in finding a suitable market before his funds were exhausted. An all-important factor in these cases appeared to be the personal goal of the inventor. If he wanted a small, easily managed organization, his company tended to remain that way. If he desired pronounced growth, was able to delegate responsibility, and was willing to work sufficiently long hours, his company grew proportionately.

Fusion Systems Corporation

In the late 1950s initial experiments were being made in producing fusion energy. The latter is the source of the sun's power. Various means for heating hydrogen and containing it in small laboratory devices were being tried. If temperatures of 100 million degrees could be achieved, theory showed that hydrogen would fuse to produce helium and tremendous quantities of energy would be released. The hydrogen bomb, in which an atomic bomb is used to produce the heating, proved that the theory was valid.

One of the heating methods that was tried in government laboratories is microwave excitation — in essence the use of a power source similar to that employed in the now-popular microwave oven. The method failed to raise hydrogen to a sufficiently high temperature, but the heated gas emitted large amounts of ultraviolet light. Five young scientists working on the project realized that they had accidentally discovered a new and superior ultraviolet (UV) source. Four of the five — Michael Ury, Marshall Greenblatt, Bernard Eastland and Leslie Levine — worked for various federal laboratories. The fifth, Donald Spero, was on the faculty of the University of Maryland. They had known each other in college. They decided, because of their discovery, to form a company to develop and market a new type of UV lamp. In 1971 they organized Fusion Systems with $10,000 of their savings. Spero resigned from his university job to become the company's president and only full-time employee. They started their business in a rented room that measured only eight by fourteen feet and was so jammed with equipment (including two arc welders) that Spero needed to sit on the sink to make notes. The first "working" model, weighing 300 pounds, was completed after four months. It failed miserably. After another year of experimentation, which included trying various lamp sizes and gas additives, a successful lamp was built. Greenblatt, who during the development phase worked full time as a computer analyst in nuclear weapons, used his spare time to contact various companies. The true market for the product proved as hard to find as the practical working design. It had originally been supposed by the five partners that the lamp would sell for the traditional uses of UV radiation; such as bleaching cloth, treating skin diseases, producing vitamin D, and manufacturing carbon tetrachloride. No substantial interest in the new lamp was expressed by any of the organizations working in these areas.

By this time the original funds of the company had run out and it was necessary to dip again into their own savings, to sell stock to friends and relatives, and to mortgage Greenblatt's house. The amount gained by these means totalled $64,000. During his contacts with various companies, Greenblatt discovered that UV irradiation was emerging as a means of drying the ink on printed surfaces. When solvent-based printing ink is used on a nonporous surface such as metal, the ink requires a long time to dry. Drying was being accelerated by gas heating to drive out the solvent. In recent years a new class of organic inks has been developed. These rapidly polymerize (and thus become solid) when they are exposed to ultraviolet light. The old mechanism of drying ink by evaporation is thus no longer necessary. It was in this application that Fusion Systems found its first glimmer of hope. The output of their microwave-driven lamp was three times that of the conventional mercury arc tubes that use metal electrodes to conduct power to the gas. Electrodes restrict the size of the arc produced in a closed tube; microwaves, on the other hand, heat all of the enclosed gas uniformly.

Beer brewers who employ UV-thickening inks for their cans became an early customer of the new company. They were to be followed by companies who distribute their products in plastic containers. The latter cannot tolerate the heat for the drying of solvent based inks.

Even at this point Fusion Systems was not out of financial danger. The partners set the initial sales price of the improved lamp, now down to 27 pounds with a ten-inch tube (Figs. 11-1A and 11-1B, at $2500. It was later found that the lamp cost $7500 to build so that the selling price had to be raised substantially. Fortunately the company came to the attention of the American Research and Development Corporation, a division of Textron, Inc. ARD is a leading venture capital firm that provides young, struggling companies with financial backing in return for an equity interest. In late 1973 ARD put $700,000 into Fusion Systems and received a fifty-six percent interest. A major break in sales came when Spero went skiing in Colorado and decided to call on a large beer brewer there. He received an order for a test model. After the latter operated successfully for twenty-one days, the brewery ordered $250,000 worth of the lamps. Fusion Systems was now established on firm ground. In 1975 the company sold eighty lamps. By the end of 1980 this figure had increased to six hundred and Fusion Systems grossed $3 million.

The company now employs fifty full-time people including two of the original partners, Spero who is still president and Ury who is vice president of engineering and development.

The founders have expressed their surprise at the difficulty of the undertaking, the long hours of work required, and the number of

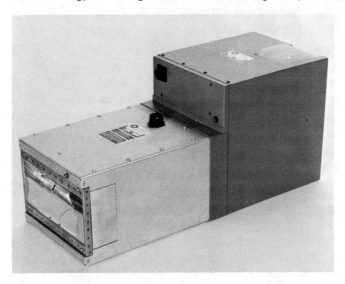

Fig. 11-1A. Ultraviolet radiator with eliptical reflector and bulb. *(Fusion Systems Corp.)*

Fig. 11-1B. Ultraviolet bulb. *(Fusion Systems Corp.)*

years needed before a net profit was achieved. Spero has summed up his experience in three separate statements.

1. "Our breakthrough was in seeing the experiment (the attempt to heat hydrogen to cause fusion) as an ultraviolet success instead of a fusion failure."
2. "We should have looked at the market first to see what people needed, rather than developing the product and then finding the best market."
3. "Ten years ago all this, (the marketing, administrative, and financial aspects of business), would have seemed boring to me. Today I can't think of anything I find more challenging."

Correlation Music Industries

Ralph G. Cromleigh is a systems designer and logic expert who has worked for aerospace companies on such exotic projects as data compression in satellite communication networks and the improvement of computer circuits. In addition to his technical interests, he has always liked piano playing and singing. In 1971 he found himself unemployed along with thousands of other scientists and engineers because of the huge layoffs in the aerospace industry.

One day, during his idle period, he was toying with a harmonica. He suddenly conceived an improved musical notation system. The idea involved the simplification of the symbols used in music and modifying the standard musical staff. Initial tests with the harmonica proved very encouraging. If certain basic changes were made in the classic system (which is of course hundreds of years old), the step between seeing a note and reproducing it on an instrument could be made much simpler. A beginner would not need to commit so many steps to memory and would thus learn much faster. Cromleigh modified a standard piano keyboard by attaching strips of electrician's tape to some of the keys. He then explained the system to his seven-year-old daughter who had never had any lessons. When presented with some music that he had transcribed into his notation, the girl learned to play it well in a very short time. Cromleigh called his patent attorney the same day.

The new system is based on the use of six lines to the staff instead of five. The keyboard is divided into six "keysets" numbered from a central zero to plus three and minus three (Fig. 11-2A). A keyset corresponds to an octave, and all keysets read the same in terms of the

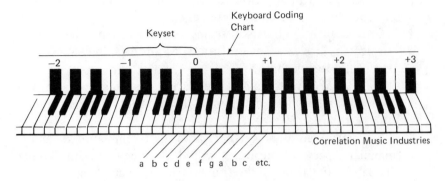

(A) Keyboard and Keysets

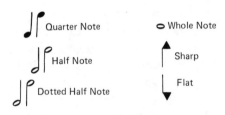

(B) Some Symbols Used in the New Notation

Fig. 11-2. The Cromleigh notation system.

keys to be struck. Ideally, the piano keyboard would be constructed with the keysets permanently marked. To allow the system to be used with present keyboards, Cromleigh devised a printed keyboard coding chart that can be inserted behind the keys. In conventional notation there are 15 key signatures and no two octaves read the same. This has been changed in the new notation. Some of the symbols used in the Cromleigh system are shown in Fig. 11-2B. A sheet of music transposed into new notation is shown in Fig. 11-2C.

Cromleigh has taught beginners to play simple melodies within five minutes.

The inventor decided that he would try to market his idea in the form of a mail-order course that included instruction manuals and a keyboard chart. However, he was unable to finance the project and still meet his living expenses. Venture capital proved impossible to find. Piano companies, musicians, and teachers were not receptive to

The Star-Spangled Banner

Fig. 11-2C. Music transposed into simplified notation. *(Correlation Music Industries)*

the idea because it represented a drastic change. By 1973 he found some consulting work with engineering companies and started to recover financially. He located an accomplished musician and teacher, a Dr. Lowndes Maury, who believed in the system and undertook to write the course. Cromleigh's basic patent was granted; he then applied for others on the system. By 1975 he was working full time again as a systems designer and was able to start his business on a moonlighting basis. His first advertisements appeared in the classified section of major magazines. By analysis of his sales he was able to eliminate those publications that did not draw responses. He found that a two-step advertisement (i.e., an ad in which the reader is asked to send for information) was better than attempting to sell directly. His ratio of sales to inquiries was one to three, a very respectable performance. Encouraged by this, Cromleigh experimented with display ads. Although these were much more expensive, the cost per response was the same. The display advertisements drew more information requests, increasing the total sales volume. The most effective ads were those depicting a cartoon drawing of a piano player with a caption that was identical to the classified ad he had previously used. Interestingly, the best magazines for his purpose were not music-related. The tabloids that are sold in many supermarkets now account for a major portion of his sales.

Cromleigh maintains accurate records and graphs of the results of each advertisement. When the cost per inquiry for a particular publication begins to increase or gross sales figures sag, he starts to consider other magazines. This droop in performance can be caused by a lessening of reader response, an increase in advertising rates, or a drop in magazine sales.

Cromleigh also gives lectures to various groups and has appeared on television. During his lectures he demonstrates the effectiveness of the system by teaching a randomly selected member of the audience to play simple tunes within a short time. He generally obtains a fairly large number of orders after one of these appearances.

The course is available in three lengths, the shortest costing nineteen dollars and the longest forty dollars. The material is printed up and stored in Cromleigh's home, which also serves as an office. A college student is employed for four hours per day. The student manages three others who each work six hours per week. These employees stuff envelopes, assemble courses and key charts, and carry out most of the other office procedures. Placing of advertisements, ordering of printing, and handling special inquiries are then taken care of by Cromleigh in approximately four hours per week. His largest expenses are: $200 per week for labor, $600 per week for advertising, and $700 per week for printing and postage. He presently sells 6000 courses a year and grosses $150,000.

Cromleigh is still employed full time as a system engineer. He is considering expanding the course to include stringed instruments such as the guitar. If he does this, he will need to devote full time to the business.

Dave Benedict Crossbows

In 1956 David Benedict was working as a test supervisor in the quality assurance department of a large aerospace company. He had studied engineering part time and had slowly accumulated credits over a nine-year period. One Saturday morning he took his family to a sports and recreation show in a nearby city. As he wandered about the various displays, Benedict happened onto an exhibit that would change his life. An English manufacturer of archery equipment was displaying crossbows and had set up a small shooting range. Up to this point Benedict's only interest in weapons stemmed from his training and experiences in the Marine Corps. When he tried the crossbow, he found its performance to be disappointing. The arrow speed was unimpressive and the precision from shot to shot was poor. The next day and in the following weeks he kept thinking about ways to improve crossbows. In his evenings Benedict designed and began to build what he considered to be a superior weapon. As he was completing the bow in his garage workshop, a curious neighbor asked to try it. Impressed, the neighbor next wanted to know what Benedict would take for it. Not particularly desiring to part with the bow, Benedict quoted what was intended to be an impossibly high figure, $110. The quote was accepted on the spot. The bow did perform much better than anything commercially available at the time. When archery enthusiasts saw the neighbor using the bow on local ranges, Benedict soon obtained as many orders as he could fill in his small shop. The word spread and business soon exceeded his manufacturing capacity. He had now come to the point that some inventors dread: Should he drop a safe livelihood and go into a promising but uncertain venture, or should he start the business on a moonlighting basis? His available capital was too small to permit the purchase of machine tools, and he was unable to find lenders. Moonlighting would mean too much time away from his family. Benedict decided to compromise. He worked half-time for various engineering companies and devoted the rest of his time to building crossbows. His wife returned to work in order to help with the family income.

As Benedict's business and reputation grew, he was able to rent space in an industrial park and buy power equipment. After seven years of part time operation, he resigned his job and went full time into the manufacturing business — a step he has never regretted.

By this time he had worked out additional ideas on how to improve the bow even more. He was fascinated by the project to a degree he had never before experienced in any of his previous jobs.

The crossbow was invented by the Chinese in 400 B.C. It has been used in every war since that time, in industry, and in sport. It was outlawed in medieval times as an "unholy weapon," except for use against heathens. More recently it was banned by some states in this country as being unfair to game. When cocked and loaded, the bow is in a stored energy state like a rifle. This permits the user to rest the weapon on a firm surface and to aim carefully before pulling the trigger. Upon firing there is very little reactive "kick," no explosive sound is emitted, and the arrow (called the "bolt" or in ancient times the "quarrel") is virtually noiseless in flight. Subjected to many centuries of research, it had supposedly reached its peak of development in recent times. Between the years 1600 to 1900 laminated steel springs (clamped together like automobile springs) had been adopted. Short arrows with spirally grooved tails were developed to impart spin stabilization during flight. These bows had a range of 1200 feet, and at shorter distances the arrow could penetrate a one-inch plank. The spring, however, required a windlass, or special lever, to permit the user to exert the one thousand or more pounds of force needed for cocking.

It seemed to Benedict that the efficiency of these bows was very low. For one thing, the spring was mounted in the stock and below the path of the arrow. The string thus pulls downward as well as forward along the stock as the weapon is fired. He found that the friction thus created accounts for as much as forty percent of the spring's stored energy. Benedict made his springs out of two fiberglass laminated sections and joined them by a push-in metal holder that is perforated to allow passage of the arrow. The metal holder is mounted above the stock. The string thus travels parallel with the stock and slightly above it (Fig. 11-3A). As a result of this change, the Benedict crossbow needs only a pull of 140 pounds to give the same performance as its predecessors did with one thousand or more pounds.

Other improvements by Benedict were based on studies made with high-speed motion pictures and electronic timers. He chose fiberglass over steel as the spring material because his pictures indicated that steel springs released energy too rapidly for the arrow to absorb efficiently and then convert into forward motion. The excess was dissipated as wave motion and irregular vibration in the bow string. Other innovations include a highly reliable safety device that must be depressed (Fig. 11-3B, page 216) before firing, arrow hold-downs to allow a hunter freedom of movement without the possibility of dropping the arrow, a safe and reliable string release,

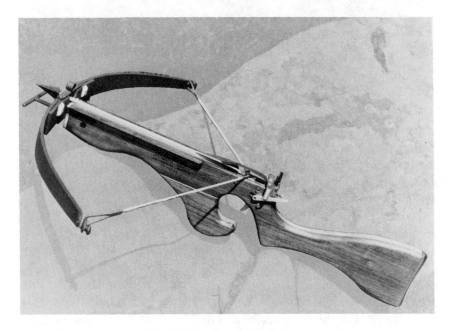

Fig. 11-3A. Benedict crossbow.

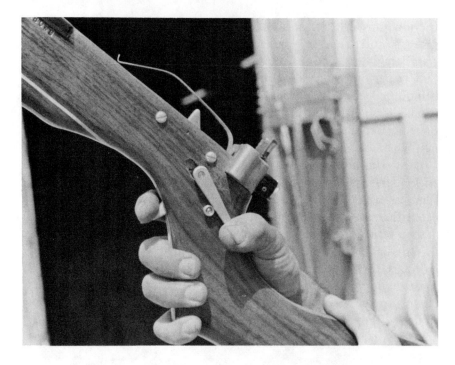

Fig. 11-3B. Crossbow safety lever in depressed position.

and a smoothly operating trigger mechanism. These and further improvements have been patented. In the present form the bow drives an arrow at a speed of 300 feet per second and will shoot an arrow inside a two-inch circle at 50 yards.

Benedict uses display advertisements in magazines devoted to hunting, archery, and guns. He also sells to dealers and sporting goods stores. His crossbows are regarded by weapons experts as the finest ever built.

He has been approached over the last fifteen years by many large companies as a takeover possibility but has resisted this because he fears that large-scale production would compromise the quality of the product.

He utilizes local shops for stamping and precision casting of the parts in the release mechanism, but manufactures the laminated stocks and bow springs himself. A bow sells for about $550, and he only manufactures about two hundred units a year. Arrows and accessories (such as telescopic sights) provide additional income. He has a long waiting list including customers in every part of the world. His bows have been used to down game as large as elk and elephant. The bows are also used by power companies to throw lines across wide gullies, by movie studios for special effects, and by the armed forces against sentries in war time.

Benedict normally employs three helpers on a part-time basis but assures the fine construction and accuracy of each bow by doing much of the work himself. His records and paper work are handled by family members or by one of the helpers. He will at times stop production to test out a new idea that promises to improve the bow even further. Benedict also manufactures a crossbow in pistol form, but this is done on special order.

In this case the inventor has maintained essentially complete and personal control of all phases of the operation – research, development, manufacturing, marketing and distribution. This has required a herculean effort and has limited his income, but his personal requirements and interests have been well served.

Gravity Dynamics Corporation

Many inventions have been concerned with chemical compounds, mechanical devices, and other methods designed for the relief of pain. The problem in selling these inventions has not been a lack of potential customers, but convincing various medical and, in some cases, government authorities that the compound, device, or method is effective and safe. In numerous instances the invention originated as a result of the inventor's personal need. This was the case with Victor Steele who suffered a back injury during boyhood while lifting weights. This ailment became chronic. During his college years he devoted his spare time to the study of bone structure and orthopedics in the hope that he would learn of some cure. After college he found work in the insurance business; he was amazed at the rise in claims that were based on back injuries. Increased interest in sports, greater automobile usage, and more exposure to industrial machinery could account for this accelerated onset of back problems. Very little relief was being achieved by classical orthopedic methods. From his studies Steele began to suspect that the upright position in which man carries himself was responsible for a slow or zero healing rate when the lower back is injured. Traction devices that involve suspension by the heels or from straps under the arms have been used for years to attempt to straighten the spine and to reverse disc compression that is a frequent source of pain. It occurred to Steele that known devices for traction were ineffective because the spine's shape is such that the vertical force through the body's center of gravity when it is suspended by the heels or arms produces further compression in the discs and even induces shearing in the disc structure. This increases the pain. In addition, traction devices make no provision for exercise, which is an important part of restoring muscular strength and resilience in the back.

One day while watching some apes in the Bronx Zoo, Steele noticed that they partially resolved the pull of gravity on their bodies by balancing upside down on a trapeze bar with the line of support at the pelvic area and not behind the knees as would be supposed. With pelvic support the spine is stretched and straightened. Gravity pull (as far as arm, shoulder, and twisting motion is concerned) is now confined to the upper part of the body so that exercises can be done with much less exertion. Excited by these findings Steele at first tried to duplicate the animal's position on a trapeze. Although the position was uncomfortable, he experienced for the first time considerable relief from his back pain. When the president of the insurance firm (himself a sufferer from lower back problems) heard of Steele's ideas, he became interested and helped finance the research.

By 1960 Steele had devised a limited motion, padded trapeze that stretched and straightened the back. The principle on which the apparatus is based is shown in Fig. 11-4A. The support members of the trapeze are curved. When the user is in position, there is no net rotational force on his body; his feet rest comfortably on the curved members. The complete apparatus is shown in Fig. 11-4B, page 220.

The individual using the system remains suspended for several minutes twice a day. He then begins to perform exercises recommended for his particular condition. He can flex his back by grasping the frame of the machine or twist his body in a series of simple calisthenics to strengthen back muscles.

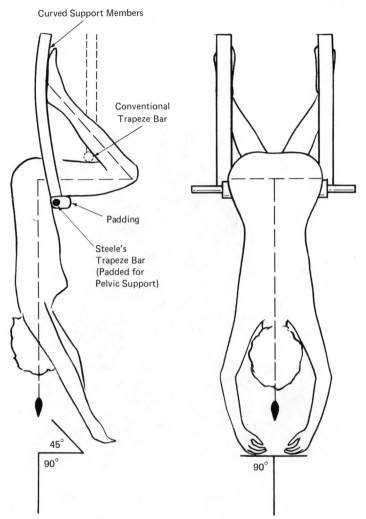

Fig. 11-4A. Back straightener.

Steele patented the machine and several of its components. His research problems were then replaced by financial ones; he could obtain no loan for manufacturing the machines; he built several of them by hand and tried to sell them to people he saw leaving the offices of orthopedic doctors. The Los Angeles Athletic Club bought one of the machines on a trial basis in 1963. It was so successful that several members ordered them for home use. It was, however, a long,

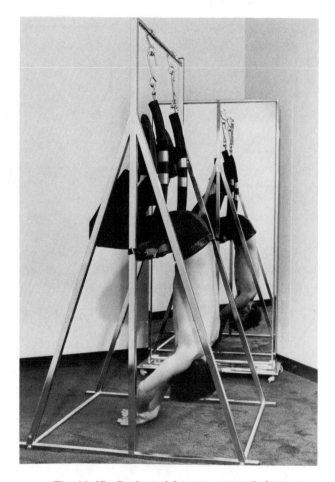

Fig. 11-4B. Back straightener — overall view.

slow, and uphill climb that lasted for twelve years. Steele supported himself in the meantime by a series of part-time jobs. His office and plant were housed in a rented garage. Gradually his volume increased from one machine every two months to his present sales and rental rate of more than thirty per week.

In the early years Steele made good use of his free time in the long intervals between sales by introducing his machine to medical schools and hospitals. He convinced doctors to experiment with them and to conduct biomechanical studies. The good results were his best advertisement. He gained considerable support from physicians who began to refer patients to him for equipment and wrote prescriptions for treatment with his apparatus.

Steele's sale of machines includes consulting and exercise planning. In cooperation with the patient's doctor, he devises special programs to handle most effectively the individual's problems. Results have been obtained in alleviating the pain of chronic lumbar conditions, herniated discs, sciatica, injuries, and other back disorders.

Steele sells three models of the machine; a doorway version, a portable model that folds and can be carried in a specially made bag, and the free standing model. Prices range from $675 to $1050. The machines can be rented for $125 to $175 per month with an option to buy.

The Gravity Dynamics Corporation, which Steele heads, employs six people to handle assembly of parts (manufactured and plated by local shops), inventory control, and office procedures. The company's gross income is $500,000 per year. In addition to physicians' referrals, Steele uses display advertising in the sports sections of newspapers.

12

Legislative Changes and the Inventor

Shortcomings of the System

Even though the United States patent system has unquestionably encouraged the highest degree of technological development ever achieved in the history of man, there are many who feel that the present legal structure has basic shortcomings which lead to the unfair treatment of the inventor and the public. It is not surprising that any body of laws which has been little changed over a long period of time would begin to show need for updating. It can also be seen that a duality of purpose will arise from any legislation which seeks to encourage one group of citizens; those not in the favored group will press for and obtain laws which equalize the privileges created for the others. The government is thus caught in the situation of a parent who must not show favoritism to any one of many children. The ambivalence of the government's position in regard to the patent law structure comes about from the concept of the monopoly afforded by a patent grant, on the one hand, and the body of laws which forbid monopolistic practices, on the other. An operating zone must be defined which allows the inventor (or the corporation which manufacturers his development) to work in an area of restricted competition, yet does not unfairly restrict trade.

Other problems with the patent system have come about as the result of the phenomenal growth of technology itself. As inventions are developed and placed on the market, there arise needs and opportunities for more inventions. The original automobiles, for example, were fairly complex when compared to the horse-drawn vehicles they replaced. The wide acceptance of the automobile encouraged a host of inventions concerned with carburetion and ignition systems, cylinder boring methods, mass production devices, rubber tire fabrication, etc. As these various inventions were incorporated in manufacturing and provided improved automobiles, still greater numbers of inventions were forthcoming. The differentiating of patent rights and the establishment of working rules connected with legal matters, such as interference, infringement, validity (and even the defining of what does and what does not constitute a patentable development) became highly important. It was inevitable that such rapid growth would produce a somewhat chaotic state of affairs.

It has been the view of many individuals that the patent law system should be "reformed." Unfortunately, it is difficult to change any long established legal structure without introducing many unforeseeable maladjustments and inequities. Many studies have been made and many new patent bills have been introduced into various legislative bodies, but none of these bills has as yet become law.

The following would seem to be the major shortcomings of the patent system as seen by its critics:

1. The increasing number of patents which are granted on marginal improvements or devices offering limited technical benefits. These patented ideas seldom go into production but they "clutter up" the patent literature. A large number of patents are granted on simple variations of existing technology. It is often easy for the competition to overcome these patents by "design-around" procedures or by the elimination of one or more elements; the money spent by the patent holder is therefore not efficiently utilized for his protection.

2. The constantly changing classification system made necessary by the advances in technology. These changes must be made rapidly in order for the system to handle the high load. It is possible, as a result, for prior art to be overlooked until after the patent is granted and the owner attempts to sue for infringement. The discovery of the prior art then invalidates the patent.

3. The time delay between the application and the·issuance of a patent. The delay is often beneficial to the applicant or corporation because it is essentially a period of protection during which no revelation must be made of technical details. Critics feel, however, that the general public is denied the benefits of this knowledge during the delay period.

4. The cost of obtaining and litigating a patent. It is felt that the financial burden is unfair to the independent inventor and places him at a disadvantage when he is competing with the development departments of large corporations. Infringement and interference actions are long and costly. Independent inventors have been unable to defend their patents against infringement by corporations who maintain large legal staffs.

5. The misuse of patents by certain groups who wish to keep a development off the market or maintain artificial price differentials between patented and unpatented products. Groups of companies will pool patents in order to drive out competition. Some individuals accumulate large numbers of marginal patents in order to pre-establish a "nuisance" position in an expanding field so that they can later obtain high royalty payments from others who make successful breakthroughs. Some companies use the patent right to gouge the consumer. It is estimated that two drug patents alone — the antibiotics tetracycline and ampicillin — made it possible for their holders to overcharge the public by more than two billion dollars. Up to the present time, the manufacturers of tetracycline have paid about 250 million dollars in settlement of treble damage suits.

6. The length of the exclusive right granted by a patent is inflexible. Many feel that a patent right should be terminated if the owner fails to manufacture or license the development within some period less than the life span of the patent.

7. The rights of inventors who work for a corporation are often traded away in documents which these individuals are forced to sign as a condition of employment.

8. The patent system of the United Stated is often incompatible with that of foreign countries so that an inventor or assignee

with overseas patents finds himself with widely varied sets of rights. It is sometimes difficult for him to prevent infringement because the device can be fabricated in a country in which he has limited or no rights and can then be imported into the United States.

Proposed Legislative Changes

Changes in patent law were made in 1836, 1870, and 1952. These were concerned primarily with codifying the body of laws and decisions which had accumulated in the system. It is to be noted that any set of laws which is actively used to regulate a human activity will not remain completely static. Challenges and variances and the resulting judgments of courts will continually set up precedents from which interpretations on new cases can be written. It is then necessary to group and condense these measures in order to eliminate contradictory or unjust regulations. The overall body of laws will, however, remain the same and will, in time, prove insufficient to handle changing conditions.

In 1958, Congress began an intensive examination of the patent law structure. In 1965, President Lyndon Johnson established the Presidential Commission on the Patent System. As a result of these lengthy studies, 35 recommendations were made to the President and Congress. Since 1967, over 15 bills dealing with patent law reform have been written and submitted to Congress for consideration. Some of these have gone from committee to full vote. As of the present writing, no bill has passed both Houses of Congress, but successive versions have approached closer and closer to passage. Perhaps the most comprehensive bill of those proposed was the one presented by Senators McClellan, Burdick, Hart, and Scott to the Committee on the Judiciary on February 24, 1976. This bill incorporated some of the prominent items of the earlier proposals which, in turn, reflected the recommendations of the Presidential Commission, opinions of expert groups such as the American Bar Association, and the results of numerous hearings. It will be profitable to examine the major features of their proposal. The changes from the present law, which would have had the most direct effect on the independent inventor, include the following:

1. The term of a patent would be extended to 20 years from the present term of 17 years. It would however, be necessary for a holder to pay "maintenance" fees after an initial period of seven years. On the seventh anniversary, the fee would amount to $300. On the tenth anniversary, another fee of $600 would be due. On the thirteenth anniversary, the sum of $1,100 would be paid in order that the patent right be retained to maturity. This change was meant to discourage "patent right collecting" and to remove non-profitable or non-exploited items from the protected status afforded by a patent. Payment of the seventh anniversary maintenance fee by an independent inventor could be deferred to the time that the tenth was due and both be paid together if he proved that his total profit up to the seventh anniversary was less than $300. A related change involved the start of the term. The patent's life would start from the date of the application and not from the date of issuance, as is now the case. The maintenance fee concept has drawn fire from inventors' groups, but experience with this change in the law will be necessary to determine whether the overall result would be harmful or beneficial to individual incentive.

2. The total fee for the application, examination, and issuance would be restricted to a maximum of $100 (less if examination is deferred, see item 7, p. 227) for the individual inventor or small businessman. The fee would be $200 for all others. Present fees start at $65 for the application and increase with the number of claims. At issuance, the applicant must pay $100 plus additional charges, depending on the number of pages and the number of drawings.

3. An applicant for a patent would be required to file a memorandum citing prior art relating to the patentability of his invention. The memorandum must explain why he considers the subject matter of his application patentable in view of the prior art. The use of these memorandums might reduce the work of the examiner and speed up the initial examination process.

4. After issuance of a patent and within 12 months under the proposed legislation, any person may submit information to the Commissioner of Patents concerning its validity. The inventor

is then notified that he has 60 days to submit a brief defending any or all of his claims. If the Commissioner feels that there is any basis for declaring the patent invalid, the matter is turned over to a board made up of Examiners-in-Chief. The patent owner may present new or amended claims during this proceeding. The ruling takes place within the Patent Office and court costs are thus avoided. It would still be possible, however, for either party to appeal the decision to the U.S. Court of Customs and Patent Appeals or to institute civil action in the U.S. District Court of the District of Columbia.

5. One year after issuance, any person could request a reexamination of a patent and submit examples of prior art. The inventor or assignee would be notified and could defend his claims by submitting a brief to the Commissioner of Patents.

6. A patent might be applied for by an inventor or by the purchaser of rights to the invention. At present, only the inventor is able to seek a patent. This feature would make U.S. Patent Law similar to that of the other countries.

7. Examination of an application would be deferred unless the applicant specifically asked for immediate examination and payed the extra fee (maximum of $100 for an individual inventor). The period of deferment would allow the inventor to determine what commercial value his innovation would have before he payed for its examination. He must, however, ask for the examination within a period of five years or his application would stand abandoned. The deferred examination concept is used in West Germany, Japan, and Australia; it serves to reduce the backlog of inventions waiting to be processed. If the examination were deferred, the contents of the application would be published after 18 months. This would not affect the inventor's rights in the matter but would inform the public of his discovery.

8. An international patent procedure would be partly set up in accordance with the Patent Cooperation Treaty signed in Washington on June 19, 1970. Rules of the treaty define the treatment of patent rights in various countries and attempt to reach an international standard. Not all features of the treaty have been ratified by the United States at this time, however, so

that this part of the bill does not provide for the complete processing of international patents. The overall aim of providing multi-nation protection with a minimum number of documents is a definite step forward in patent reform. A number of difficulties concerning certain secret matters (e.g., atomic weaponry or defense patents) would definitely need clarification before this highly commendable effort could be fully implemented.

9. Fraudulent and deceptive treatment of inventors by unregistered individuals would be made a federal offense which could be punished by fines up to $10,000 and imprisonment up to one year. Presently, the actions of certain development companies in misrepresenting their functions in obtaining patents are difficult to punish under federal law, and remedy must be found in civil action or under state laws (which vary widely). Registered attorneys and agents are held to ethical behavior and the maintaining of good faith by existing legislation; the latter part of the present law would not be changed. The inventor himself is required to deal in candor and good faith with the Patent Office.

10. A United States patent could be considered infringed under special conditions if an item were manufactured abroad by a process covered by the patent. This provision would cover the following cases:

    a. The person who manufactures abroad in order to avoid paying royalties on a valid U.S. patent and then imports the product into the United States.

    b. The person who has an exclusive sales license in the United States for such a manufacturer.

    c. The person who manufactures components in the United States and then exports these for assembly into a foreign country knowing that such an assembly in this country would be subject to civil action.

The above list represents only some of the salient features of the bill but gives some idea of presently considered changes. It appears likely that most of the above items will appear in the legislation which is finally adopted.

A sore point for many years has been the patent rights document signed by individuals who work for others. The employee, as a condition of being hired, signs over the rights to any invention he makes and agrees to make no demand for compensation other than the salary he is paid. To properly evaluate this practice, it is necessary to consider the various conditions under which an employee can or will produce an invention. The following are a few:

1. The employee is a qualified scientist or engineer who has been hired to solve corporate problems. He is being paid to do research which is ordinarily expected to lead to inventions. In a sense, this case is analogous to that of a miner who is engaged to find and remove minerals occurring on the property of another. Presumably, the research worker will be paid even if his answers to problems are not patentable.

2. The above employee does an adequate job for the corporation but also has hobby interests in an unrelated field. The employee invents several profitable items on his own time and with his own materials, but the agreement he signed states that all inventions made by him after his starting date belong to the employer.

3. The employee, in this case, is not a trained technician and works in a non-production job in a factory (e.g., accounting, maintenance, payroll, etc.). On the basis of a casual observation while walking through the plant, he invents a labor-saving process of great value to the company. He did not sign a patent rights agreement when he was hired. The employer feels that he has "shop rights" in the invention because the employee used information available to him only as a result of the job.

4. Another non-technical employee invents a device which is of no direct interest to the corporation but it can be shown that he did part of the development work on company premises after hours, using company materials.

5. An engineering consulting company contracts to solve a problem for an agency of the federal government. One of the engineers produces a valuable invention during the course of

the work. Does the patent right belong to the contracting firm, to the inventor, or to the government?

6. An employee leaves one company to go to work for a competitor. He carries in his head certain information which leads to a valuable patent. The patent is assigned to the second company, which pays him a large bonus. The first company claims the rights on the basis of an agreement he signed with them.

In the absence of a national policy on patent ownership questions several states have undertaken to clarify the matter within their own borders. In the last few years there have been introduced into four state legislatures bills that have features aimed at achieving fair treatment for employed inventors. The first to become state law is Minnesota's Freedom to Create Act. This went into effect on January 1, 1978. The law defines two kinds of invention; the "free invention" and the "service invention."

For it to be classified as "free," an invention must meet the following criteria:

1. It is not directly related to the employer's business or anticipated research and development;
2. It is not the result of any work performed by the employee for his employer;
3. It has not been developed in any part using the employer's time, equipment, or trade secrets.

A "service" invention does not fall under any of these criteria. The law states that a free invention belongs to the employee; a service invention is owned by the employer. Any pre-employment agreement signed by an individual, which donates rights to free inventions, is void and unenforcible.

A similar law was enacted in California and went into effect January 1, 1980. The California law also distinguishes between free and service inventions and protects the employee's rights in the former. The burden of proof in any lawsuit arising from the new legislation is, however, on the employee. Two other states, Washington and North Carolina, have since passed similar laws.

These four laws are significant steps towards improving the situation of the employed inventor but are only a beginning. Because the measures apply solely within the areas of the four states, large companies and conglomerates may have little difficulty in sidestepping them. It would also be difficult to define the range of an employer's business in the case of a multicompany conglomerate. Almost any discovery by an employee could be defined as being within the company's interest.

There have been efforts by various professional societies, notably the Institute of Electrical and Electronic Engineers, to draft and obtain sponsorship for a bill in Congress that would result in a United States law instituting reforms similar to those in the four states listed above.

European Patent Organization

Pressure for patent reform has not been confined to the United States. Dissatisfaction with the complexity and cost of obtaining multinational patent protection has, after twenty years of effort, resulted in the form of a group called The European Patent Organization. Ten European countries have signed an agreement to set up a unified patent-granting assembly.

For an inventor to patent his discovery in more than one country, he is normally required to apply separately in each. He needs to obtain legal representation in the countries involved, to pay each set of fees separately (and these vary greatly from country to country), and to have his application translated into each appropriate language. A patent protects only in the country of issue. Persons in other countries may copy the invention without infringing on the patent (but cannot by international law bring the finished goods into the country of issue).

Under the new arrangement, it is possible to file a single application and to obtain a patent in two or more of the countries enrolled in the European Patent Organization. These countries are at present England, France, West Germany, Italy, Belgium, Netherlands, Luxembourg, Sweden, Austria, and Switzerland. It appears likely that Denmark and Ireland will join in the future.

An application may be submitted in any one of three "official" languages: English, German, or French. The patent seeker may elect to apply in as many of the member countries as he wishes with a single application. The cost at present is approximately $2500 for all the countries. To this must be added one attorney's or patent agent's charges and various fees for each country. This totals to more than the cost for any one country but is considerably less than it would be if done separately. The total cost for patent coverage is relatively low when the area and potential market (300 million persons) is taken into account. This is particularly advantageous for outsiders such as citizens of the United States.

The EPO, headquartered in Munich, will eventually be manned by 2000 employees, most of whom will be drawn from the national patent offices of the member countries. The EPO also has a search department in the Hague, where applications will be studied and compared with the technical literature and records of prior art.

Time will determine the effectiveness of the EPO. Although increased protection at lower cost would certainly provide correspondingly greater motivation for the inventor, there would seem to be some complications for which no suitable provisions have been made. Infringements and interferences would be settled under the provisions of the country or countries where they occurred. This might involve hiring several attorneys. Another complication is that of language differences. Specifications and claims in any language are often finely shaded in order to sidestep existing patent bars such as prior art, obviousness, etc. If the application or issued patent is challenged in another country using both the latter's translation and sets of laws, it is possible that a very difficult-to-resolve legal situation might result. Still another problem concerns the concept of maintaining the confidentiality of the contents of the application while the patent is being examined. In some countries this is not the practice, while in others the application is published. Revealing the contents prior to allowance in country A would destroy the significance of keeping it secret in country B.

Recent Court Decisions

In Chapter 1 we mentioned a few examples of items that could not be patented. Among these were naturally occurring substances — even if these have not been discovered by anyone before. In a recent decision (Diamond vs. Chakrabarty) the U.S. Supreme Court ruled that a strain of bacteria developed through genetic crossbreeding was patentable. Chakrabarty, a microbiologist, had evolved a new strain of bacteria to attack oil spills and thus help clean up seawater after a maritime accident. The court found that the bacteria were not a naturally occurring substance but had been "created" in the same sense that a composition of matter is formed by human effort.

The patent laws also state that mathematical formula and methods are unpatentable. Thus computer programs and software in general have been ruled out. The Patent Office initially rejected an application called "Digital Control of Rubber Molding Presses" on the basis that the only novel feature of the system was the use of computer control. The latter employed a mathematical formula to set the press temperature for producing a proper cure of the rubber. The Supreme Court, on the other hand, held that the combination of press, computer and software "solved a practical problem in the molding of rubber products." This decision will release for patenting a large number of applications involving software, but it is still likely that software alone (i.e., not in combination with a control process) may be found unpatentable.

In the above discussion, we have attempted to present an impartial view of present thinking on patent law reform. The problems attendant on rewarding accomplishment, preventing abuse, granting limited monopoly, and retaining incentive, all the while holding the general public's welfare in mind, are numerous and difficult. It has required the carefully directed efforts of many legislators and specialists to achieve the results so far obtained.

Index

Adaption of new principles
   to an old problem, 18
   to new uses, 48
Adhesives, 85
Advertising, 195, 212
Aluminum-epoxy composition, 162–165
Analysis and synthesis, 87–93, 111–113
Angular position indicator, 138–140
Antibiotics, 224
Apparatus
   crudeness vs effectiveness of, 91
   elegance in, 136–137
Application, abandonment of, 189
Applying for a patent
   costs, 179–181
   documentation, 182
   procedure, 182
   reasons for, 175–177
Area principle, 32
Areas of invention, 13
Assignment of the patent, 190
Attorneys and agents, 178, 179
   working with, 181, 182
Automatic volume control, 122
Automobiles, 223
Averages, 164

Back straightener, 217–221
   illustrated, 219, 220
Benedict, Dave, 213–217

Bias vs impartiality, 161, 178
Biological modeling, 84–86, 105
Brokers, 195–196

Camera rangefinder, 106
   illustrated, 107, 108
Centrifugal pumps, 109
Changing viewpoints method, 80, 107, 109
Chromatograph, 137
Claims, 45, 53, 54
   rejection of, 183
Class and subclass of patent, 46, 65–66
Compensation of apparatus
   in measurements, 158
   null methods for, 158–160
   optical, 160
Compositions of matter 4, 7, 9, 128
Consulting by inventors, 189, 202
Continuation patent, 185
Conveyors, 99, 100
Creativity enhancement, 73, 75–94
   changing viewpoints, 80–92
   listing methods, 75–97
   matrix methods, 75–80
   question asking, 82–83
   updating and adaptation, 83
Cromleigh, Ralph G., 209
Crossbows, 213, 217
   history of, 214
   illustrated, 215, 216
Cross fertilization, 92, 113–114

Data
    mathematical treatment of, 163-165
    uses of, 162
Dehydration of gypsum, 128-131
Dempster, A.J., 136
Design of measuring systems, 157-162
Design patent, 4, 9, 175
Development companies, 177, 196, 197
Diligence and follow up, 97, 172
Direct solution, 17
Disclosure forms
    corporate, 187
    proposed revisions, 229-231
Divisional patent, 185
Dosage, 134
Drawing
    as a means of modeling, 138
    perspective, 138-140
    schematic, 138-140
    working, 138-140

Edison, Thomas, 83
Electric razor, 200
Electrodeless UV lamps, 207
    illustrated, 208
Electroplating, 5, 6
Employed inventors, 171, 229
Energy
    conversion of, 22-23
    flow of, 27-29
    forms of, 22-24
    storage of, 29-30
    transfer of, 27
Equilibrium principles, 36
European Patent Organization, 231-232
Experiment planning
    purposes, 115-116
    steps in, 117
Expired patents, 169

Fabrication processes
    for hard materials, 142
    for soft goods, 142-143
Feedback principles, 33
Fees, 179, 181, 184
Fire escape apparatus
    analysis of, 88-89
    improvement of, 89

Flow
    laws, 28
    of energy, 27-29
    of material, 27-29
Flywheels, 29
Fog penetration, system, 94
Foreign patents, 224, 231
Friction, 24
Function reversal, 80, 107, 109
Functional diagrams, 117
    symbols for, 118
Fusion energy, 205, 206
Fusion Systems Corporation, 205-209

Gasoline turbines, 71-72
Gazette, U.S. Patent, 66, 194
Glass-filled nylon resin, 104
Golf ball retrievers, 46-56
Golf tee, 187
Gordon, William J.J., 94
Group methods, 93-95
    brainstorming, 93
    cross stimulation, 95

Heart monitoring, 81, 82
Hilsch tube, 21
Historical studies,
    use of patents in, 57
Holography, 113-114
Honey, 84
Human memory, 113-114

Impedance matching, 157
Impingement pump, 107
Index of classification, 63, 64
Individual inventor, 2
Infringement, 192
Inner conflicts, 170
Instrument characteristics
    dead zone, 150-151
    noise, 151-154
    range, 148, 149
    sensitivity, 149
    speed of response, 151
Inventing
    as problem solving, 70-71
    commercial aspects of, 169
    full or part time, 168

"grazing" aspects of, 168
job security and, 171
Inventing
    subconscious processes of, 171
    teaching of, 3
Inventing methods
    problem solving and, 71
    study of, 70
Invention
    belief in, 186, 187
    definition of, 4
    definition of problem, 73
    kinds of, 14, 70
    manufacturing of, 198-201
    marketing, 186
    plants, shrubs, etc., 4, 9, 175
    to order, 168
    what is not, 9, 10, 11, 175
Inventors' clubs, 197, 198
Inventor's rights, 229-231

"Jury rigged" experiments, 124

Kubie, Lawrence S., 170

Labor saving concepts, 15
Large and small principle, 33
Large scale machinery, 201
Lasers, 113-114
Length standards, 148
Libraries containing patents, 67
Licensing, 192
    restrictions, 192
    sub-licenses, 192
Lifeboat antenna
    experiment plan, 123
    experimental design of, 119-123
    height, 119
    purpose, 118
    for receiving, 118
    for transmitting, 121
Light emitting diodes, 116
Locating buyers, 193
    advertising, 193, 195
    mailing list methods, 193
    trade shows, 195
Lotion, 9, 131-133

Machines, 6
Mail order marketing, 195, 212

Manufacturing the invention
    advantages, 198
    degree of involvement, 199
    packaging, 200
    pricing, 200
    systems for, 199
Marine propellor, 19, 20
Marketing services, 195-196
Martin, A.J.P., 136-137
Mass spectrograph, 136
Matrix methods, 78-80
Measurements
    general types, 147-149
    maintaining impartiality, 160-162
    transducers for, 147
"Mental jogging", 97
Method of manufacturing, 4, 7, 8
Mixing, alloying and adjoining, 34-36
Modelmaking
    error minimization, 145-146
    general principles, 143-146
    optimization techniques in, 144
    simplifying, 143
Models, types of, 140, 141
Monopoly rights
    implied by a patent, 222
    "nuisance" value of, 223
Motorized drum, 99
    illustrated, 101
Motivation to invent, 167
    causing specialization, 170
    inner needs determine, 170
Moveable core transformers, 154
Musical notation systems, 209-210

Naturally occurring substances, 11, 232-233
Navigation of bees, 85
Noise
    filters for, 153-154
    in measurements, 151-153
Novelty, 9, 170

Official Gazette, 66
Old principle, applied to old problem, 17, 18
One piece racquet, 102, 104-105
    illustrated, 105
Optical amplification
    noise in, 153
    for pressure measurement, 152-153

Order and chaos
    "absolute zero", 31
    expanding universe, 31
    tendency to breakdown of various
        systems, 32
Ore sizing, 124-129
Osborne, Alexander, 93

Particle flight, 124
Patent
    advisers, 178-179
    applied-for stage, 183
    class and subclass, 46, 65-66
    costs, 179-181
    development companies, 178, 179
    for a golf ball retriever, 46-56
    generic statements in, 52
    Official Gazette and the, 66, 194
    search, 46, 63-67, 69
    services, 177, 178
Patent system shortcomings, 223-225
    proposed changes, 225-228
Patentability, 9-13
    and diligence, 10
    and prior art, 110, 175
    and public well being, 10
    inventor-created bars to, 176
    obvious solutions and, 12-13, 175
    of methods of doing business, 11
Patented inventions
    marketing of, 189, 190
    pricing of, 191
Patents
    abstracts of, 44
    as technical literature, 56
    claims in, 45, 53
    drawings for, 46, 51, 180
    history of the U.S. system of, 42, 43
    number of, 43
    references in, 45
    shortcomings in the U.S. system of,
        222-225
    structure of, 43-56
    title blocks of, 43
    using to trace technical development, 57
Pencil leads using plastic binder, 18
Penicillin, 84
Performance clause, 191

Perpetual motion, 24-26
    living organisms and, 26, 27
    water falls and, 25
Pharmaceutical inventions, 132-135
Photosynthesis, 30
Pitot tube, 110
Placebos, 133
Plaster of paris, 129
    tests of, 131
Playback equalization, 59
Plug-in switch, 100-102
    illustrated, 103
Poison ivy lotion, 132, 133-134
Poulsen, Valdemar, 57
Power of attorney, 182
Practice table of combinations, 96, 97
"Predoomed" approaches, 21
Prefabricated materials, sources of, 143
Prestressing, 37, 38
Prior art, 1
Problem definition, 73, 74
Processes, 4
Product-by-process claiming, 135
Product improvement, 74
Projection of particles, use of conveyor in,
    125
Proposed revisions, U.S. Patent Laws,
    225-228
Push pedal bicycle, 111-113
    illustrated, 112

Quality control principle, 34
Quick opening valve, 91-93

Racquets, 102, 104, 105
Radio transmission of power, 83, 84
Randomization, 161, 164
Raw material forming
    methods for, 142
Recent court decisions, 232-233
Reduction to practice
    working models for, 25, 26
"Reinventing", 12, 95, 173
Reissue patents, 185
Royalties, 191

Saline solutions, 121
Satellite power stations, 201

Saturation principle, 37
Scientific principles and mathematical formulas
    patentability of, 11, 233
Scope of invention, 202, 203
Searching of patents, 46, 63–67, 69, 175
    effectiveness of, 69
    libraries for, 67–68
    revealing prior art by, 173
Self-generating principle, 38–39
    generating of spheres, 39
    military rafts, 39–41
    self-sealing, 39
Separation of powders, 60–62
    handbook explanation, 60
    patent for, 62
Separation processes, 126, 127
    efficiency of, 127
Serendipity, 19
Shielding, 146
Short term inventions, 76
Single or multiple combinations, 14, 95, 99
    mail order catalogue method for, 15
Solar heaters, 79–80
Specialization, 13, 14
Specialized instruments, 202, 203
Speedboat, 141, 142
"Spin-off", 171
Spring hinges, 144
Standard deviation, 164
State laws protecting inventors, 230–231
Steel "whiskers", 33, 34
Steele, Victor, 217–221
Steinbeck, John, 94
Subliminal learning, 173
Surplus equipment, use of, 143

Switches, 100, 102
System development, 75

Tape recorder development, 57–59
Technological growth, 222, 223
    and patent classifications, 223
Thermocouples, 153
Time measurement, 156, 157
Tires, 80
Transducers, 154–157
Typewriter eraser, 188, 189

Ultrasonic ranging, 105–107
Ultraviolet lamps, 206–209
Underrating principle, 34
Unwarranted secrecy, 170, 181
Updating and adaptation, 83
Urban transportation, 111
U.S. Patent Office
    backlogs and delays in, 181
    "Rules of Practice", 180
UV thickening inks, 207

Vibration, 146
    measurement of, 158
Voltmeters, 121, 148–151
    "dead zone" in, 150–151
    scales for, 148, 149
    sensitivity of, 149
    speed of response of, 151

Warm up methods, 95
Wire recorders, 57–59
    Camras, Marvin, 59
    problems in, 58, 59
Working models, 141
Writing pen, 141